# WEDGWOOD

# WEDGWOOD

Geoffrey Wills

**SPRING
BOOKS**

*Frontispiece*
Josiah Wedgwood (1730–95) painted
by George Stubbs in fired colours on a biscuit
earthenware plaque. Inscribed 'George Stubbs Pinxit 1780';
50.8 × 40.6 cm. (20 × 16 in.). Wedgwood Museum,
Barlaston.


First published in Great Britain in 1980 by
Country Life Books and distributed for them by
The Hamlyn Publishing Group Limited

This edition published in 1988 by Spring Books
An imprint of Octopus Publishing Group PLC
59 Grosvenor Street
London W1

Distributed by The Hamlyn Publishing Group Limited
Bridge House, London Road,
Twickenham, Middlesex, England

Copyright © Geoffrey Wills 1980

ISBN 0 600 55729 4

Printed in Hong Kong

# Contents

# Foreword

ALL WHO write on Wedgwood owe a big debt to Eliza Meteyard. It was she who penned the first important book on the subject; a book that is still indispensable after more than one hundred years.

Joseph Mayer, who was born at Newcastle-under-Lyme but made his home and his fortune at Liverpool, allowed Miss Meteyard full access to his great hoard of Wedgwood documents. These he had found in 1848 under what might be considered unpropitious circumstances: caught in a sudden downpour when in a Birmingham suburb, he took shelter in a rag-shop where he had the incredible luck to discover the majority of the Wedgwood records that had been disposed of following the death of Josiah II in 1843. In the intervening five years, much of the accumulation had been sold to local shopkeepers as wrapping-paper, for as the ragman is stated to have explained, '. . . folks fancy their bits o' butter and bacon all the better if wrapped in clean writin'.' In due course the papers rescued by Mayer came into the possession of Josiah Wedgwood & Sons who, in turn, have deposited them at the University of Keele.

In addition to being grateful for the pioneering work of Eliza Meteyard, the present writer acknowledges with thanks various forms of assistance rendered by the undermentioned:

the many who have contributed to the pages of the *Proceedings* of the Wedgwood Society; Alison Kelly, Editor of the *Proceedings*, for the loan and use of the manuscript of her article on the Catherine service; J. K. des Fontaines, Chairman of the Wedgwood Society, for kindly allowing use of his researches into Wedgwood bone china; H. L. Douch, Curator of the County Museum, Truro, for drawing attention to an advertisement in the *Sherborne Mercury*; and for unstinted aid; Gaye Blake-Roberts and Lynn Miller, respectively Curator and Information Officer at the Wedgwood Museum, Barlaston; and Derek Halfpenny, Public Relations Director of Josiah Wedgwood & Sons.

1. Staffordshire agate-ware tea caddy and cover, *c.* 1750; height 14 cm. (5½ in.).
City Museum & Art Gallery, Stoke-on-Trent.

# CHAPTER ONE

# Early Days

IN THE NORTHERN part of the county of Staffordshire two materials led to the founding of an industry – an abundance of clay that is suitable for making pottery and the coal with which to feed the kilns for baking it. This industry began by supplying local requirements, but by the outset of the 18th century was sending many of its products farther afield. In the course of time, the area became known as 'the Potteries', a name that remains to the present day.

The numerous individual potteries or 'pot-banks' were small concerns, each with a kiln or 'hovel' on land adjoining the dwelling, and almost all of them worked by their owner with the aid of a few employees, often including some relatives. It is necessary to discuss the earlier products and the methods of their manufacture in order to appreciate the advances due to the enterprise of Josiah Wedgwood.

The traditional goods of the district were for a long time of a coarse type, mostly made in a heavily glazed red clay and ornamented in a variety of ways that demanded limited skill. The composition of the glaze included a proportion of lead oxide and, unavoidably, impurities in the form of iron that gave it a yellow/brown cast. Any white clay ornamentation therefore acquired that colour, and articles made from red and white clays left the kiln in a combination of shades of reddish-brown and yellow. The majority of the objects made were dishes and pots of all sizes and shapes and, although some of them were intended as decoration in the home, most were severely functional.

Some new and better finished types of ware were introduced in the late 17th century, and these soon superseded the older and less sophisticated kind. In one instance, the red clay was employed after having been treated for the removal of impurities to make it of a fine texture, and then fired in a considerably hotter kiln than had been customary. This ware was first produced in Staffordshire by two brothers, John Philip and David Elers, who came from Holland and worked in the county for a few years from 1693. The red pottery they made was finished without a glaze, the great heat of the kiln having vitrified the clay making it watertight. The teapots and other objects attributed to these two men were either quite plain and smooth, or ornamented with relief patterns.

Another improvement has traditionally been attributed to a local

2. Staffordshire teapot and cover, *c.* 1750, of the unglazed red stoneware introduced in the 1690s by the Elers brothers. Impressed imitation Chinese seal. Width 17.2 cm. (6¾ in.). City Museum & Art Gallery, Stoke-on-Trent.

potter, John Astbury, who died in 1743. He is said to have been the first to import and use white clay from Devonshire, with which he made contrasting reliefs on his redware. Astbury, like the Elers brothers, made his ornaments with metal stamps, moistening the dried clay object and sticking each relief in position. This simple method of 'sprigging' was improved by the use of moulds made of unglazed pottery or plaster, which gave much better results. Soft clay was pressed into the mould, carefully removed and then affixed as before.

Some of Astbury's contemporaries made articles of red, locally found white, and stained clays semi-blended to form more-or-less realistic imitations of natural stones and marbles. These were known as 'agate' wares which, being of the same substance throughout, are to be distinguished from another variety named 'marbled'. The latter was similarly patterned, but only on the surface.

A further innovation in Staffordshire was salt-glazed stoneware, which had long been known in Germany but was first made in England by John Dwight of Fulham only in the third quarter of the 17th century. It was manufactured from white clay that had been rendered more

3. Astbury-type teapot and cover of red clay with white reliefs covered in glaze. Staffordshire, c. 1750; height 13.3 cm. (5¼ in.). Victoria and Albert Museum, London.

manageable when plastic, and both more durable and white when fired, by the inclusion of powdered flints. The hard stones were calcined: that is, they were heated until red-hot, cooled, and ground to produce a fine white powder to be mixed with Devonshire or other clay before being fired at a high temperature. However, fragments excavated on the sites of Staffordshire potteries have shown that much of the ware was made from a coarse near-white clay mixed with sand. Before firing, the articles were each dipped in a locally dug pipeclay slip so that they were completely covered. The finished ware differed very little in appearance from that made of clay and flint throughout, and the deception is usually only revealed by chipping or breakage. The use of cheaper ingredients found in the neighbourhood kept down the cost and the selling price of 'dip-ware', and it continued to be made for several decades.

The Germans discovered that a decorative glaze could be given to such ware by the use of common salt, which was shovelled into the kiln when the contents were at red heat. The introduction of the salt resulted in the emission of fumes from the kiln, and Simeon Shaw wrote of 'the vast volumes of dense clouds of vapour proceeding from the ovens at

4. Staffordshire salt-glazed stoneware sauceboat, *c.* 1750; length 19.7 cm. (7¾ in.). City Museum & Art Gallery, Plymouth.

the time of employing Salt for the purpose of causing the glazed inside and outside of the Pottery'. He added that because of Burslem's elevated situation the clouds dispersed in a few hours, but some fifty years later L. M. Solon wrote that the thickest London fog was a fair comparison with the state of the atmosphere when glazing took place.

A mixture of white clay, local or imported, together with flints began to be used with a lead glaze, but fired at a similar temperature to that used for the red wares. This was done because the glaze could not withstand the heat required for stoneware, but had the advantage of being cheaper to make as it needed less fuel. In due course the lead-glazed pottery with its yellowish tint was gradually improved and became known as creamware. For a long time, the glaze was applied to the naturally dried clay article in the form of powdered galena (sulphide of lead) or lead oxide (litharge: white or red lead), a single firing completed the operation. According to tradition, *c.* 1750 Enoch Booth of Tunstall discovered that a marked improvement resulted from the use of a liquid glaze. This was applied to the object after it had first been fired to an unglazed 'biscuit' state, when it was dipped in a liquefied mixture of ground lead and flints before having a second firing.

One other ware deserves a mention: 'Black Egyptian' which was a black unglazed hard pottery, a stoneware, that was the same colour throughout. According to one writer it gained its name from its general resemblance to classical Greek and Roman black pottery, but may have been so-called because its colour recalled the dark skins and jet black hair of gypsies who were sometimes known in the past as 'Egyptians'. This ware was made either from clay containing manganese in its composition or from a dark red clay to which oxide of manganese was added.

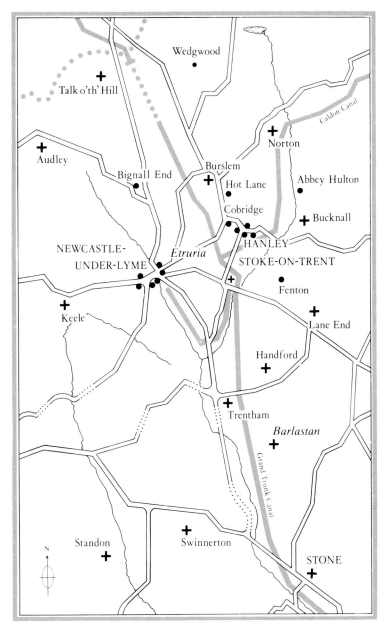

5. North Staffordshire *c*. 1800, showing places mentioned in the text.
This area is still known as 'the Potteries'.

Jugs and teapots were, according to Simeon Shaw, the principal products of makers of this type of pottery.

The Staffordshire potworks were to be found in groups, having been established wherever suitable deposits of clay existed. In time, the hamlets expanded and were linked to one another by tracks and roads. The largest of the communities was at Burslem, at some point named 'the mother of the Potteries'. It had achieved renown as early as 1686, the year when Robert Plot published his *History of Staffordshire*. He wrote:

> But the greatest pottery they have in this county is carried on at Burslem, near Newcastle-under-Lyme, where for making their different sorts of pots they have as many different sorts of clay, which they dig round about the towne, all within half a mile's distance, the best being found nearest the coale. . .

At a later date, Simeon Shaw added a little further information gathered from elderly acquaintances in the locality. He stated that the Burslem district boasted about twenty-two kilns at the close of the 17th century, adding, 'no manufacturer of that period fired more than one oven full weekly, commencing on the Thursday night, and finishing about mid-day on Saturday'.

The town shared its name with that of a family, one of whom, Margaret Burslem, in 1612 married Gilbert Wedgwood who had settled in the area. Other Wedgwoods, direct or indirect descendants of Gilbert, owned potteries in the neighbourhood of Burslem and elsewhere, the Churchyard pottery, Burslem, being from the mid-17th century owned by successive bearers of the name Thomas Wedgwood.

Thomas Wedgwood III, of the Churchyard, was born in 1687 and duly married Mary Stringer, daughter of a Unitarian minister of Newcastle-under-Lyme. According to John Ward, the historian of Stoke, their first child, Margaret, was followed over the years by a dozen others: five girls and seven boys, the last being a boy they named Josiah. Other writers have stated that Josiah was the youngest of ten children, and others again have made him the youngest of twelve, Josiah C. Wedgwood, writing in 1908, subscribing to the last number. Similarly disputable is the exact date of his birth, but it is clearly recorded that he was baptised at Burslem on 12 July 1730. Notwithstanding these and other uncertainties regarding his early years, he was destined to transform the local pottery trade into an internationally recognised industry; one with which his own name, and that of his descendants, is inseparably linked.

Considerable imagination has been exercised at various times in describing how the young Josiah may have spent his childhood, successive writers indulging in the maximum of deduction from the few recorded facts. There is no evidence to support the assertion of Llewellynn Jewitt that 'he was an amiable, thoughtful and particularly intelligent child, ever quiet and studious, and delighting more in thoughtful occupations than in the games and rough exercise of the boys of that, and indeed of every, time'. Eliza Meteyard suggested that he would have been a pupil at a dame-school in Burslem, 'more to be out of the way of mischief, than for the learning to be obtained there', although it is uncertain whether such an establishment existed in the place. When he was seven years of age it would appear that Josiah began to attend a school at Newcastle-under-Lyme, 'kept by a man of superior education named John or Thomas Blunt'. According to another tradition, the boy was adept at using a pair of scissors, apparently making what were later known as silhouettes of somewhat unusual subjects. These were said to have included

... an army at combat, a fleet at sea, a house and garden, or a whole pot-work, and the shapes of the ware made in it. These cuttings when wetted were stuck the whole length of the sloping desks, to the exquisite delight of the scholars, but often to the great wrath of the severe pedagogue.

In the first half of the 18th century, as in later years, the majority of people, apart from the aristocracy and their connexions, lived their lives and left behind them little more than the dates of their baptism, marriage and death. Many thousands came and went without even those statistics. It is improbable that there would survive any reliable records, if there ever had been any, relating to the most junior of a country-based family of no especially great wealth or other distinction.

One occurrence that must have affected Josiah's youth would have been the death of his father in 1739, when Thomas Wedgwood was 52 and his youngest child under nine years of age. The Churchyard pottery that had been inherited by Thomas was bequeathed by him to his eldest son, his namesake and the fourth of the family to become its owner. Each of the six younger children, who included Josiah, was to receive the sum of £20 on reaching the age of twenty, 'and that in paying the said sums, the elder shall be preferred before the younger'. The executors were charged with using any money 'not settled on my Wife' for the upbringing of the younger children, and despite this it has been stated that Josiah immediately ceased to attend school and went to work under his brother at the Churchyard.

Whether he remained at school or not, it is known that he had the misfortune to contract smallpox when the disease made an appearance at Burslem, but the year in which this took place remains debatable. Some say 1741 and others 1746 or 1747. In any event, it would seem that the attack was a serious one, and after a period of convalescence Josiah found that his right knee remained so stiff as to seriously impede movement. The condition eased with time, but it was reported that thenceforward he required a stick when walking.

In chronological order the next certain fact is that on 11 November 1744 when he was fourteen years of age, Josiah was apprenticed to his eldest brother. The Indenture of apprenticeship made it clear that

. . . the said Josiah Wedgwood, of his own free Will and Consent, and with the Consent and Direction of his said Mother, Hath put and doth hereby Bind himselfe Apprentice unto the said Thomas Wedgwood, to Learn his Art, Mistery, Occupation, or Imployment of Throwing, and Handleing, which the said Thomas Wedgwood now useth . . .

In other respects the document contained the usual stern admonitions against playing dice and cards, haunting ale-houses, contracting matrimony, and so forth.

In the following year the fifteen-year-old youth would undoubtedly have been aware of the local excitement caused by the presence in the vicinity of the Young Pretender's troops and the English army under the command of the Duke of Cumberland. A few of the former penetrated as far south as Talk o' th' Hill, only five miles north-west of Newcastle-under-Lyme. A Shelton man wrote at the time of the occurrence reporting that the Pretender's men:

. . . took all young Breck's horses, and drank him a deal of liquor, but did no other mischief; and our armie lay encamped at Stone Town field with their artillery and every thing in very great order, which was such a sight as was never seen by any body in our country.

Josiah reputedly began his apprenticeship by learning how to throw, the art of making and shaping hollow vessels on the potter's wheel. The flat wheel, or circular board, was placed horizontally at a convenient working level, and an axle through its centre connected to another wheel at the base, this latter being made to revolve by the action of the worker's foot. There can be little doubt that the damage to his knee would have made it difficult for Wedgwood to operate the device, and it is assumed that he was forced to transfer his attention to other sides of the craft. It cannot be confirmed, however, that he was forced, thereafter, 'to sit whilst at work with his leg extended before him upon a stool.'

When his time with his brother terminated in 1752, Wedgwood entered into partnership with two men named John Harrison and Thomas Alders, at Cliff Bank, Stoke. Harrison was described by Jewitt as 'a man possessed of some means, but little taste', and he apparently provided capital for the venture. The output is said to have included stoneware articles, either plain or of the variety known as 'scratched blue'; the latter being decorated with patterns incised in the semi-soft clay which were dusted with blue for emphasis.

6. Staffordshire salt-glazed stoneware mug with 'scratched blue' decoration, dated 1752; height 12.7 cm. (5 in.). Victoria and Albert Museum, London.

Many of the stoneware goods would have been produced by press-moulding, or by the recently introduced process of slip-casting. In the former, a thin slab of soft clay was placed between two parts of a metal or alabaster mould, and then squeezed so as to receive the shape of the mould and the impression of any relief patterning in it. Slip-casting took advantage of the porous nature of plaster of Paris, of which patterned moulds were made. A quantity of liquefied clay, or 'slip' as it is called, was poured into the mould and after a few minutes was tipped out leaving a thin coating of clay within due to the absorption of water by the mould. The operation would be repeated until the layers formed a sufficient thickness and it was then allowed to dry and harden enough for the mould, which was made in close-fitting parts, to be removed. The plaster soon lost its sharp details, but was easily replaced as often as required by fresh moulds cast from a master 'block' of hard pottery. The undecorated surfaces of articles made by the foregoing methods vary in appearance, and provide clues as to which was employed. Press-moulding leaves the area flat and smooth, while slip-casting results in the reverse side showing depressions where there are reliefs in the pattern.

By 1754 the Alders–Harrison–Wedgwood partnership broke up, and the latter entered into an agreement with one of the foremost potters of the period, Thomas Whieldon, of Fenton. Whieldon made a wide variety of goods, and from surviving documents it is clear that he was in business on what was then a large scale. Excavations carried out on the site of his principal pottery at Little Fenton, just to the north of Stoke-on-Trent, resulted in the discovery of quantities of fragments that included red clay wares, salt-glazed stoneware, and varieties of creamware. Many of the latter bore mottled glazes of the type known as 'tortoiseshell', because of a faint resemblance to that material, which has always been associated with Whieldon's name.

Articles of the above types were being made widely in Staffordshire by 1750, where there was an increasing awareness of competition from porcelain. This was more expensive to buy, whether imported from the mainland of Europe, the Far East or from one of the newly started English factories. Whatever its source, it set standards of quality in both appearance and durability that fully justified the cost for those who could afford it. Above all, its whiteness and attractive decoration had great appeal, which was especially merited in the case of plates and dishes where porcelain contrasted so favourably with the less appealing and less hygenic pewter and wood.

Among Josiah Wedgwood's fellow-workers at Thomas Whieldon's pottery was an apprentice, William Greatbatch, son of Daniel Greatbatch, a local carrier and farmer, who became adept at devising new shapes and patterns. He worked as a designer for Whieldon, and when Wedgwood left in 1759 so did Greatbatch. It was Greatbatch who was responsible for designing the teapots and other articles in the form of fruit, made in creamware and coated in clear glazes, that are usually attributed to the years when Wedgwood and Whieldon were at Fenton. These pieces were the pottery-makers' attempt to compete with the comparable fashionable vegetable objects made in porcelain at nearby Longton Hall, and with the animals, birds and fishes, formed as tureens, produced at Chelsea.

When he was at Fenton, Wedgwood made many efforts to improve existing wares, and an unusual clause in his partnership agreement permitted him to keep to himself any discoveries he made: he was under no obligation to reveal them to his partner or to anyone else. It was at this time that he perfected a clear and bright green glaze that had been known in Tudor times and subsequently neglected. His report of the successful trial respecting the colour was recorded by him in these words: 'A Green glaze, to be laid on Common white (or cream color) biscuit ware. Very good–March 23, 1759.'

Wedgwood's private notebook, in which are listed the above and other experiments, has survived, and is in the Wedgwood Museum at Barlaston. He prefaced it with an interesting survey of the output of the pottery at which he worked and a general picture of the state of the industry, drafted at the time and subsequently amplified:

7, 8. Dovecote ornament and cream jug of creamware with mottled coloured glazes, *c.* 1755; heights 12.4 cm. (4⅞ in.) and 13.3 cm. (5¼ in.). City Museum & Art Gallery, Stoke-on-Trent.

This suite of experiments was begun at Fenton hall, in the parish of Stoke upon Trent, about the beginning of the year 1759, in my partnership with Mr. Whieldon, for the improvement of our manufacture of earthenware, which at that time stood in great need of it, the demand for our goods decreasing daily, and the trade universally complained of as being bad & in a declining condition.

White Stone ware (viz. with Salt glaze) was the principal article of our manufacture. But this had been made a long time, and the prices were now reduced so low that the potters could not afford to bestow much expense upon it, or make it so good in any respect as the ware would otherwise admit of. And with regard to Elegance of form, that was an object very little attended to.

The next article in consequence to the Stoneware was an imitation of Tortoiseshell. But as no improvement has been made in this branch for several years, the country was grown weary of it; and though the price had been lowered from time to time, in order to increase the sale, the expedient did not answer, and something new was wanted, to give a little spirit to the business.

I had already made an imitation of Agate; which was esteemed beautiful & a considerable improvement; but people were surfeited with wares of these variegated colors. These considerations induced me to try for some more solid improvements as well in the *Body*, as the *Glazes*, the *Colours*, & the *Forms*, of the articles of our manufacture.

I saw the field was spacious, and the soil so good, as to promise an ample recompense to any one who should labour diligently in its cultivation.

The final paragraph of Wedgwood's remarks was prophetic, and within a short time of writing those words he began the task of tilling the 'spacious field'.

# CHAPTER TWO

# The Bell Works

FOLLOWING THE termination of his partnership with Whieldon at
Fenton, Josiah Wedgwood returned to his birthplace. Earlier, he had
made plans to set up in business on his own account, for there survives
an agreement he made with his cousin, Thomas Wedgwood. The document is dated at Stoke, 30 December, 1758, and by its terms Thomas
'engageth to serve the s^d. Josiah Wedgwood as a Journeyman from the
first of May 1759'. Eliza Meteyard, who printed the Agreement in full,
drew attention to the several dates for the start of Josiah's venture given
by various writers; they ranged between 1756 and 1760, but it seems
without doubt that the intended day was 1 May 1759. There is also some
argument about the precise location of his first premises, Jewitt stating
that he began by occupying the family's Churchyard pottery, but
although he may have been elsewhere briefly at first it is agreed that he
leased the Ivy House pot-bank. The owners were again relatives, John
and Thomas Wedgwood, and for the kilns with the adjoining creeper-
clad house in the centre of Burslem they charged him £10 a year.

Simeon Shaw gave the details he had managed to gather regarding
the buildings, their situation with regard to convenient taverns, and the
kinds of goods produced.

Mr. Josiah Wedgwood returned to Burslem, about 1760, and commenced
Business alone, at the small manufactory (at that time *thatched*, as usual,) to be
seen from the bar of the Leopard Inn; very near that of his distant relations,
Messrs. T. and J. Wedgwood, and only a short distance from that of his father.
Here he continued the manufacture of Knife Hafts, Green Tiles, Tortoiseshell
and Marble Plates, glazed with lead ore, for his previously formed connections;
and his attention to their demands soon secured him such a share of business,
that he engaged a second small manufactory, only across the high road, and
where is the Turk's Head tavern. Here he manufactured the White Stone pottery,
then increasing in demand; and there yet remain of this kind, white Tiles, with
relief figures, of a Heron fishing, and a Spewing-Duck fountain.

Within a couple of years Wedgwood had outgrown the Ivy House
and took the lease of another Burslem pottery: the Brick House. It later
became known as the Bell Works, because Josiah built on it a cupola in
which hung a bell. With this his employees were summoned to work

each day, a novelty that performed its task while exciting general interest in an area where the blowing of a horn had hitherto achieved the same end. The installation and use of the bell can be seen as an early example of Wedgwood's flair for publicity, and his continual careful regard for small matters that might have wide effects.

He must have been aware instinctively that the fragmented Staffordshire industry could not compete much longer with cheaper and better imported goods, either pottery or porcelain. Despite the imposition of import duties, the public's appetite for them was seemingly insatiable and they continued to enter the country in large quantities. Porcelain, especially, was coming from China in an unending flood, and there was apparently no limit to the amount that could be made there. It is not unlikely that Wedgwood knew of the letters explaining the details of porcelain-making sent back from the Far East by the French Jesuit missionary, Père d'Entrecolles. The letters had been printed in France soon after they were written, in 1712 and 1722, and again in a description of China by J.–B. du Halde that was first issued in Paris in 1735 and published in an English translation in 1738–41. D'Entrecolles described with great care the materials and processes in use at Ching-tê Chên, where there were some 3,000 kilns and a population of one million employed in the industry in one capacity or another. 'I am told,' he wrote, 'that a piece of porcelain has passed through the hands of seventy workmen.' He lamented the fact that among the men employed in kneading clay by treading it, his converts to Christianity found difficulty in attending church: '. . . they are only allowed to go if they offer substitutes, because as soon as the work is interrupted all the other workmen are stopped.'

This example of industrialisation was extreme, and Wedgwood's milder imitation of it, introduced gradually, became an important factor in his success. The other key components shared by East and West were the achievement of an output of consistently good quality and a strong desire to please the buyer. His years from 1759 to 1765, in particular, were devoted to these ends and set the standards for future decades.

Soon after he became established on his own, Wedgwood was doing business with his old acquaintance, William Greatbatch. A note from the latter dated 22 July 1760 specifies a number of items 'Left at the Cross Keys, Wood Street, London', an inn used by coaches with destinations as far apart as Hertford, Hoddesdon and Ware to the north of the city, Wantage to the west, and Norwich to the north-east. The goods included 'foxglove teapots', 'melon sauce boats and stands', and 'Woodbine' and 'Chinese' teapots. Examples of some of these and of the other patterns listed have been identified in stoneware and creamware.

The range of pieces formed as fruits, mentioned earlier as having been

9. Wedgwood-Whieldon teapot and cover in the form of a pineapple, *c.* 1755; height 10.8 cm. (4¼ in.). County Museum, Truro.

made in the Whieldon–Wedgwood era, included pineapples and vegetables, which were similarly attractively glazed. A note from Greatbatch dated May 1764 reads, in part:

There are ready two crates of Pine Apple ware. . . . The order of the Pine Apple-ware w$^{ch}$. Thomas gave me will be compleated the next Ovenfull – and his order consist of about 80 Doz. should be glad what sorts to make to compleat the order you gave.

In other letters there is evidence that Greatbatch was making blocks (*see* p. 17) for Wedgwood as well as supplying him with teapots and other items. Most, if not all, were unglazed and were given their finish by Wedgwood, and while some were of fine creamware throughout others were of a coarse clay that was disguised by the glaze. All are notable for clearcut modelling and their strong, pure colour.

Josiah was making his own creamware by 1761, if not a little earlier. Like that being made elsewhere, it was of a deep cream, almost buff, tone, and at first with a glaze that sometimes showed crazing: a network of fine cracks. It was to be some years before he was able to perfect a truly cream-colour earthenware with a near-colourless and even glaze.

There is evidence that in the early 1760s Wedgwood was no less busy as a buyer and a salesman than as a potter. Extant bills show that from 1762 he was engaged in buying pottery from several local makers;

10. Wedgwood-Whieldon teapot and cover in the form of a cauliflower, *c.* 1755; height 12 cm. (4¾ in.). County Museum, Truro.

11. Block moulds for two teapot spouts and for a cauliflower teapot like that in the preceeding illustration, all probably designed by William Greatbatch. Height of teapot 12 cm. (4¼ in.). Wedgwood Museum, Barlaston.

equally he was selling to them, but as none of the ware bore marks it is only possible to hazard a guess as to the source of surviving examples. However, a feature pointing to a Wedgwood origin is to be found on his teapots; in many instances, Donald Towner gives the proportion among existing examples as being as high as 90 per cent, the spouts are of a distinctive pattern. They are moulded with overlapping cabbage or cauliflower leaves, and used indiscriminately on plain or moulded pots, of the same design as the spouts of many of his vegetable and pineapple teapots. Saltglazed stoneware blocks of such spouts are to be seen in the Wedgwood Museum, Barlaston, having been discovered in 1905 at Etruria, where they had lain unnoticed for over a century.

To interest a wider public it was essential that the pottery should have a more sophisticated appearance. Moulded patterns in relief dimly seen through daubs of transparent coloured glazes were unlikely to prove of compelling interest to buyers who admired painted porcelain. Two other styles of ornamentation were available: straightforward painting, or the newly devised process of transfer printing. Small potteries did not always have resident decorators on their staffs, and in that case they could send the partly finished goods to persons or firms specialising in the work.

As regards painting, Wedgwood is known to have had an unspecified proportion of his ware decorated locally. In one instance it was sent to a painter of whom nothing is known, his name being mentioned in a note dated 12 July 1763 and addressed to Wedgwood:

> Sir
> I shall send Mr. Courzen's ware to his painting shop tonight.
> Wm. Greatbatch.

A year or so later Josiah was beginning to have dealings with David Rhodes of Leeds, perhaps the same Rhodes who later became his employee. In 1760 the *Leeds Intelligencer* carried an advertisement for Messrs Robinson and Rhodes, announcing that they enamelled all kinds of chinaware, and 'sell a good assortment of Foreign China and great variety of useful English China of the newest improvement, which they will engage to wear as well as Foreign, and will change gratis if broke with hot water'. A letter to Wedgwood signed by Rhodes 'for Partner and Self', dated 11 March 1763, contained an order for teapots and fruit baskets and dishes, adding 'we shall want many more of them'. Without doubt these goods would have been undecorated and were painted as required by the recipients, who then sold them to their clients. There is no evidence at this date that Wedgwood was having his ware painted on his own account, and he was confining his own decorating to the application of coloured glazes in the Whieldon manner.

12. Creamware teapot and cover painted in colours by David Rhodes, *c*.1760; height 14 cm. (5½ in.).

Simeon Shaw, who usually recorded events that took place before his own time, and was in no position to argue with what inhabitants of the area said that they remembered, wrote:

> The tea ware required to be painted, was sent for that purpose to Mrs. Astbury, in Hot Lane; which was sold, packed, and sent away from Burslem; and some time elapsed before Mr. W. had the enamelling executed on his own premises.

Pieces that may have been decorated by Mrs Astbury of Hot Lane, Cobridge, have not been identified, but some have been attributed to Robinson and Rhodes. They are painted in a naive and colourful manner with figures against rustic backgrounds, flowers, or with simple geometrical patterns, all of which would probably have had a greater appeal in the country than in London. There they would have been competing with the more skilfully decorated products of Chelsea, Bow, Worcester and other English porcelain factories as well as with imported goods.

Josiah Wedgwood recognised that hand-painted work demanded skilled artists for its execution, and that such persons were difficult to find and expensive to employ. Printing as an acceptable alternative to painting became available very soon after he became established at Burslem, and Wedgwood soon took advantage of the opportunity it presented to reduce costs while maintaining quality. Transfer-printed decoration was first employed at Birmingham in *c*.1751, when an Irish

13. Dish printed by Sadler and Green with a rustic scene, *c*. 1780; impressed 'WEDGWOOD'. Diameter 35 cm. (14 in.). Victoria and Albert Museum, London.

engraver, John Brooks, applied for a patent for a process he had discovered 'by great study application and expense . . .' By 1753 enamels were being printed with designs at a factory opened for the purpose at Battersea, but three years later the concern was bankrupt and closed.

The method was apparently also discovered independently by two Liverpool men, John Sadler and Guy Green, who testified on oath in 1756 that they had successfully decorated 1,200 tiles with printed patterns in the space of six hours. They added in their statement that they had been 'upwards of seven years in finding out the method . . .' The process was equally applicable to enamels, pottery and porcelain.

Like so many other significant inventions, it was a surprisingly simple one: an engraving or woodcut was printed with special ink on thin paper, the image then being transferred to the glazed surface of the object to be decorated. Firing resulted in the design melting into the glaze and the finished article presented a smooth and shining appearance with the fine lines of the engraving clearly apparent. For Wedgwood's ware black was commonly employed, with brick-red as an alternative, both of them varying in tone and clarity according to the accuracy of the firing; the former could vary from a pure black to a dull green or a blackish-brown. From 1770 a further colour, purple, was sometimes used.

Sadler and Green worked for Wedgwood from 1761. According to Shaw, he 'employed the waggon belonging to Mr. Morris, the carrier, of Lawton, once a fortnight, to take down a load of cream colour to be

26

printed in this improved manner, by Messrs. S. & G. and return with the load previously taken for that purpose'. An invoice of 11 April 1764 shows that Wedgwood paid the Liverpool firm about £65 for decorating 1,730 pieces of ware, the list comprising 'Teapots in three sizes, Mugs in two sizes, Bowls, all sizes, Coffee Pots, Sugars, Cream Ewers, and Cups and Saucers'.

Printing, whatever the colour, gave the pottery a sophistication that enabled it to vie on artistic grounds with porcelain, while the cost of the finished articles was much less. In a particular instance of decorative treatment there was little or nothing to choose in appearance between creamware, delftware and porcelain. This was when a pattern painted or printed in black was overlaid with a wash of transparent green which allowed the basic design to show through. The concept probably originated at the Marseilles factory where a fine tin-glazed earthenware was painted in that style from the 1750s onwards, the London porcelain-decorator, James Giles, executing similar work on Worcester porcelain in c. 1770. Wedgwood's successful treatment on the same lines, using transfer-printing on creamware is seen in illus. 15, which shows a dish from a dessert service of c. 1775.

In the early 1760s people in many areas of England were becoming aware of how much the narrow pot-holed roads and tracks impeded commercial transport as well as social intercourse. The delivery of his ware to and from Liverpool and the sending of it to other parts of the

14. Cup and saucer and covered jug printed in purple by Sadler and Green, c. 1775. All impressed 'WEDGWOOD'. Diameter of saucer 12.7 cm. (5 in.). Victoria and Albert Museum, London.

country would have brought to Wedgwood's attention the deplorable state of the roads in his part of Staffordshire. The Burslem potters in general, who had a two-way traffic with pottery leaving the area and flints, salt and white clay coming in, realised that they would never attain prosperity unless action was taken. They united to petition parliament for a turnpike: a sound road that would be paid for by tolls collected from its users.

Wedgwood took a keen interest in the scheme, almost certainly in the role of spokesman for his fellow-potters, writing from London to his cousin, Thomas, who was at home, about parliamentary progress of the Bill. If it had been allowed as proposed the road would have been given not only an improved surface and have been widened, but would have been shorter. The existing route entailed a detour through Newcastle-under-Lyme that lengthened the journey by some four miles, increasing costs and losing time. The plan was to by-pass the town, but it was strongly opposed by the inhabitants of Newcastle, especially the inn-keepers there, and due to this contrariness the Bill was passed finally in 1763 in an amended form.

Whether travelling by way of the old route or the new, Wedgwood often visited Liverpool. There, not only did he see Sadler and Green about decorating, but also he had calls to make respecting export of his wares to America and the Indies. On the occasion of a visit there in the middle of 1762 he damaged his already injured knee and the local surgeon Mathew Turner, ordered him to rest. Perhaps to alleviate the boredom of his patient, Turner introduced one of his friends, Thomas Bentley, whom he brought with him to the inn where Wedgwood was detained. 'Bentley', wrote Miss Meteyard, calling hard on her imagination, 'came forward with his gallant bow and courtly manner, took the kindly proffered hand, looked into the good and strongly expressive face of the Staffordshire potter; and from this moment, this place of meeting in the Liverpool inn, these men were more than brothers.'

Bentley was a partner in the local firm of Bentley and Boardman, merchants, and had been born in Derbyshire in 1730, the same year as Wedgwood. He had a wide range of intellectual interests, enjoyed the friendship of many talented men of the north-west, and was a Unitarian: a Dissenter, whose religious beliefs were combined with a radical political allegiance. His views coincided with those of his new acquaintance, although Bentley's education was superior; he not only had some classical learning but also was able to converse in French and Italian.

As soon as he had recovered sufficiently to stand the journey, Wedgwood returned to Burslem and hard work. There is no doubt that he was still experimenting to improve his output in quality and quantity, and was still directing his attention principally to creamware. At this date it was still of a deep colour, but no mark was used, and the principal evidence as to its tint comes from a letter of rather later date when he had perfected his pale cream colour. In 1768 Wedgwood wrote about his

> ... endeavour to make it as pale as possible ... but it is impossible that any one colour, even though it were to come down from Heaven, should please every taste, and I cannot regularly make two cream-colours, a deep and a light shade, without having two works for that purpose.

The next event in Josiah's life was a domestic one: he married his cousin, Sarah, daughter of Richard Wedgwood, a cheese-merchant of Smallwood, east Cheshire. In a letter to Bentley dated 23 January 1764, the future bridegroom wrote:

> ... we are to be married on Wednesday next. On that auspicious day, think it no sin, to wash your philosophic evening pipe with a glass or two extraordinary, to hail your friend & wish him good speed into the realms of matrimony.

The Wedgwood wedding took place in the church at Astbury, Cheshire, on 25 January.

A year later, on 3 January 1765, Josiah celebrated the birth of the first of his nine children, a daughter who was named Susannah. After she had been baptised, her father wrote to his eldest brother John that 'we have now added another Christian to the family'. In the same letter he wrote:

> It gave me great pleasure to have it under your own hand that your health & spirits are good & your affairs in so promising a way of being settled agreeably & with dispatch – this gives us some hopes of having the pleasure of your company this spring which I doubt not you will find very salutary, especially as we now have got such pretty employment for you. Sukey is a fine sprightly lass, & will bear a good deal of dandleing & you can sing – lullaby Baby – whilst I rock the Cradle ...

It suggests a pretty scene of brotherly accord, and reveals that Josiah was not so preoccupied with the pottery as to be immune to the joys of parenthood.

For some time past Josiah had made use of his brother, John, who lived in the city of London, in Cateaton Street, almost opposite the Guildhall. He was conveniently situated to act as a part-time agent for his brother in Staffordshire, arranging such things as sending him supplies of gold powder for gilding, seeing engravers for the making of copperplates, and paying visits to Josiah's clients who resided in the capital.

15. Queensware dish transfer-printed in black and overpainted with transparent green, *c.* 1775; impressed 'WEDGWOOD'. Width 22.8 cm. (9 in.). City Museum & Art Gallery, Stoke-on-Trent.

It was John to whom he wrote in mid-June, 1765. He began:

Dear Brother

I'll teach you to find fault, & scold, & grumble at my not writeing, I warrant you, & as to your going to France, I do not believe I can spare you out of London this summer, if business comes in for you at this rate, for instance – An ord$^r$. from St. James's for a service of Staffordshire ware, about which I want to ask a hundred questions, & have never a mouth but yours in Town worth opening upon the subject.

The ord$^r$. came from Miss Deborah alias De$^b$. Chetwynd, sempstress, & Laundress to the Queen, to Mr. Smallw$^d$. of Newcastle, who bro$^t$. it to me (I believe because nobody else w$^d$. undertake it) & is as follows.

A complete sett of tea things, with a gold ground and raised flowers upon it in green, in the same manner of the green flowers that are raised upon the *mehons*, so it is wrote but I suppose it sho$^d$. be *melons* – The articles are 12 cups for Tea, & 12 Saucers a slop bason, sugar dish w$^{th}$. cover & stand, Teapot & stand, spoon trea, Coffeepot, 12 Coffee cups, 6 p$^r$. of hand candlesticks & 6 Mellons with leaves.

6 green fruit baskets & stands edged with gold.

In a postscript he added with understandable excitement: 'Pray put on *the best suit of Cloaths you ever had in your life*, & take the first opportunity of

16. Set of three creamware vases with engine-turned decoration, *c.* 1765;
heights 28 cm. (11 in.) and 33 cm. (13 in.).
Saltram House, Devon.

going to Court. Miss Chetwynd is Daughter to the Master of the Mint'.

It is sad to record that there remains no other evidence of this service, every single item of it having vanished over the years. Nevertheless, this order from Buckingham House (as it was then named) so fortuitously reaching Wedgwood, gave him exactly the encouragement he needed and at the right moment.

Wedgwood was spending some of his time experimenting with the lathe, a tool not unknown in potteries in its elementary form for giving pre-fired wares a smooth finish. He realised that the more complicated machine used for ornamenting ivory, wood, and metal had an untapped potential for decorating pottery. An engine-turning lathe was adapted for the work and found satisfactory, giving wares an attractive and distinctive appearance.

Many of the improvements and modifications to the lathe were executed for Wedgwood by John Wyke of Liverpool, who specialised in watchmakers' tools and included James Watt, inventor of the steam-engine, among his customers. Wyke was a friend of Bentley and, like him, a member of the Unitarian congregation at the Octagon Chapel, Temple Court, Liverpool. The building had been erected in 1763 to the plans of Joseph Finney, who combined watch-making with architecture, and who apparently supplied Wedgwood with hand-tools when Wyke was unable to do so. The original lathe devised in the middle of the 1760s is preserved at the factory, remaining usuable two centuries later.

In a letter to his brother in London, sent a month after the one previously quoted, the successful use of the lathe is mentioned. John made contact with Miss Chetwynd, and as a result Wedgwood had received permission to send some of his wares for inspection by Queen Charlotte. The relevant paragraph reads:

> I shall be very proud of the honour of sending a box of patterns to the Queen, amongst which I intend sending two setts of Vases, Creamcolour engine turn'd, and printed, for which purpose nothing could be more suitable than some copper plates I have by me.

Vases with engine-turned ornament in the shape of closely spaced wavy ribs and flutes, but without any printing on them, have been recorded in the dark-toned creamware of the period, and it is reasonable to assume that they are similar to those referred to in the letter.

Writing to Cateaton Street at the end of July or beginning of August 1765, the letter is undated, Wedgwood detailed some of his difficulties in finishing the teaset for the Queen. The gilding was proving very troublesome, and he suggested that John should ask the advice of a man named Jinks who had worked at Chelsea and was then at Bow. The letter closed with a paragraph of local news of special interest to the writer:

> Dr. Swan dined with Lord Gower this week; after dinner your Brother Josiah's Pottworks were the subject of conversation for some time, the Cream colour Table services in particular. I believe it was his Lordship said that nothing of the sort could exceed them for a fine glaze, &c.

Lord Gower had been appointed Lord Lieutenant of Staffordshire in 1755, and successively held a number of important posts. His seat, Trentham Hall, was close to Stoke-on-Trent, so that he had an interest in the Potteries and the prosperity of the area. Like his brother-in-law, the Duke of Bedford, he was a member of the Whig party, which included in its ranks many influential men with whom, in the course of time, Wedgwood became acquainted.

In March 1765 a long-mooted scheme for a canal to link Liverpool and Hull by way of north Staffordshire was revived, and again Wedgwood rallied his fellow-potters in a cause to benefit the entire district. His friend, Thomas Bentley, participated in the project, and another close friend, Dr Erasmus Darwin of Lichfield, physician and poet, also played a part; the two men were responsible for composing a pamphlet for distribution by Wedgwood. As might be expected, Lord Gower became associated with the plan and the Duke of Bridgewater, whose engineer, James Brindley, had already constructed more than one canal, was no less involved.

At the end of December Brindley and Wedgwood had dined together, and the latter attended a meeting of interested local persons at the Leopard Inn, Burslem. Josiah reported this to his brother, adding: 'Our gentlemen seem very warm in setting this matter on foot again & I could scarcely withstand the pressing solicitations I had from all present to undertake a journey or two for that purpose'. In May 1766 he was appointed treasurer by the Proprietors of the Navigation from the Trent to the Mersey, the body formed to carry through the project. In his own words, in a list of the officers he sent John: 'Jos. Wedgwood Treasurer at £ooo per ann. out of which he bears his own expences'.

Josiah had no doubts that such a canal would assist his business, and the trade of other potters, in the same way as the earlier turnpike had done. However, in addition to expediting the carriage of goods, the preliminaries with which he was so closely concerned enlarged his clientele in more than one direction. In a letter of 6 July he wrote to John:

I should have wrote to you sooner but have been waiting upon his G– the D– of Bridgewater with plans &c. respecting Inland Navigation. Mr Sparrow [solicitor, of Newcastle-under-Lyme] went along with me, we were most graciously rec$^d$. spent about 8 hours in his G–'s comp$^y$., & had all the assurances of his concurrence with our designs that we could wish. His G– gave me an ord$^r$. for the completest Table service of Cream colour that I could make . . .

Again, he wrote to Bentley in October 1765 telling of business in hand which was doubtless owed to the good offices of Lord Gower:

I have been three Days hard & close at work takeing patt$^{ns}$. from a set of French China at the Duke of Bedford's, worth at least £1,500, the most elegant things I ever saw.

The service in question, of Sèvres porcelain, had been presented to the Duchess by Louis XV in 1763 when the Duke of Bedford was British Ambassador at Paris. Each piece was richly painted with panels of flowers or exotic birds reserved on a dark blue and gilt ground, and it remains to this day at Woburn Abbey. Josiah would have been less interested in the decoration of the service than in the shapes of the numerous items, and J. V. G. Mallet has drawn attention to a creamware dish (in the Victoria and Albert Museum) closely resembling that beneath the Sèvres soup tureen. The preserves dish *(illus. 17)* which was engraved in Wedgwood's 1774 catalogue, was certainly inspired by a *confiturier* from the same source.

In the summer of 1765 Wedgwood became concerned about the

17. Preserves dish adapted from one in the Duke of Bedford's Sèvres porcelain service. An engraving from Josiah Wedgwood's creamware catalogue of 1774.

transaction of business in the capital. In August he wrote tactfully to his brother, who was apparently retired, inquiring whether he could devote more of his time to representing the Burslem manufactory:

Pray how are you for business, or schemes of pleasure? for I wo[d]. by no means break into the latter, but if it wo[d]. be consistent with both & you sho[d]. choose it I wo[d]. send you a pattern or two & a list of my Chaps [customers], and more bills if you choose such employm[t]. till a *better place* offers. I have this year sent goods to amo[t]. of ab[t]. £1,000 to London all of which is owing & I sho[d]. not care how soon I was counting some of the money.

You know I have often mention'd having a man in London the greatest part of the year shewing patterns, taking orders, settling acc[ts]. &c. &c. & as I increase my work, & throw it still more into the ornamental way I shall have the greater need of such assistance & sho[d]. be glad to have your advice upon it. Wo'd £50 a year keep such a Person in London & pay rent for 2 rooms (both back rooms & St. Giles's wo[d]. be as good as St. James's) about so much I think it might answer for me to give.

From the foregoing it is clear that the potter had determined to enter into competition with porcelain by extending his output to include vases and other purely decorative articles, while not neglecting the market for tableware. It was by no means unreasonable that he should do so, for his cream-coloured pottery supplanted the delftware which in its day had embraced both categories of goods. He foresaw that London would be the place in which to sell both vases and services, especially the former, and he was reinforced in his opinion a week later. Then, he wrote to John that he had had a visit at Burslem from the Duke of Marlborough, Lord Gower, Lord Spencer 'and others', who made some purchases and remarked that they wondered why he had not got a London warehouse 'where patt[ns]. of all the sorts I make may be seen.'

Soon afterwards two events occurred: a couple of rooms were rented in Mayfair, in Charles Street (now Carlos Place), and John left London to live in Liverpool. Someone must therefore be found to attend to the new premises and its clients, and Wedgwood sent William Cox. The fact that this man was a bookkeeper at Burslem suggests that selling goods was considered to be less important than the rendering of accounts and collecting of moneys outstanding; bad debts were a notorious feature of the business world in 18th-century London, and all too frequently led to bankruptcy.

It was in the summer of 1766 that Queen Charlotte appointed Wedgwood 'Potter to Her Majesty'. Shortly afterwards he began to refer to his pottery as 'Queensware' and his London address became 'The Queen's Arms'. At a later date he reconsidered the matter, advertising it primarily as his 'Warehouse', the reason being 'that the *Queens Arms* may not be thought to be a Tavern, as I thought in hear$^g$. it read it sounded something like it'.

During the busy time when he was engaged in assisting in the promotion of the Trent and Mersey canal, fulfilling the Queen's order and worrying about representation in London, Wedgwood was also energetically pursuing plans to expand his pottery. As early as March 1765 he told his brother about experiments he was making with a whiter body and glaze. Then, he wrote: 'I do not intend to make this ware at Burslem & am therefore laying out for an agreeable situation elsewhere.' The 'situation elsewhere' eventually proved to be an estate of 350 acres, the Ridge House estate, a few miles to the south of Burslem and just to the north-east of Newcastle-under-Lyme.

As might be expected at any period, negotiations over the purchase of the land were protracted, but in the middle of 1767 completion of the deal seemed close at hand. Wedgwood wrote to Bentley on 26 July, '. . . I hope yet to be able to build a Vase work at the latter end of this summer', but the matter was settled only in the December of that year.

Early in the summer of 1767 the two men, Wedgwood and Bentley, agreed to become partners in the new venture. The potter had been pressing for this to take place, but was well aware that it meant Bentley would have to leave Liverpool and all his friends there. Finally, on 20 May Wedgwood was able to write:

Your most acceptable letter of the 15$^{th}$. gave me the highest pleasure in seting before me a nearer prospect that I have yet had, of a union that I have long coveted, & which I do not doubt will be lasting, delightfull, & beneficial to us both, & as to the time & manner of leaving Liverpool, make it the most agreeable to yourself in every respect, & it will be perfectly so to me.

Erection of the new manufactory began under the direction of a Derby

18. Three vases and covers of creamware imitating ormolu-mounted marbles. Each marked in relief on a disc 'WEDGWOOD & BENTLEY: ETRURIA'; heights 31.7 cm. (12½ in), 43.2 cm. (17 in.) and 26.6 cm. (10½ in.).

architect and builder, Joseph Pickford, who built at the same time a house apiece for the two partners; over one hundred dwellings were also provided for employees. Wedgwood was eager that his pottery should be as up-to-date and efficient as possible, and the plans for it incorporated some of the features of the Soho engineering workshops at Birmingham that had opened in 1765. The owner of Soho, Matthew Boulton, was admired by Wedgwood who described him as 'very ingenious, philosophical, and agreeable', both men having much in common as regards their approach to industrialisation.

While his new factory was under construction, Wedgwood was far from idle, spending some of his time developing a new variety of pottery suitable for forming into vases and other decorative objects. For this, he took the local black Egyptian, and in due course produced a finely grained hard black pottery, referred to at the time as 'Etruscan'. The first successful vases made from the new material were completed and sent to Bentley at the end of August 1768, and at the same time Wedgwood was still busily experimenting with vases made from his creamware.

The creamware vases imitated in shape and decoration prototypes of stone and marble, reflecting a taste among connoisseurs for exotic natural stones, a taste that was especially prevalent in France. Foremost among French *amateurs* was the duc d'Aumont, who had acquired some ancient porphyry brought from Genoa in *c.* 1750 and then purchased more precious marbles. By the early 1770s the duke had established a workshop in Paris where the stone could be shaped and polished, and where it could be mounted elaborately in gilt bronze. To the same end, the Sèvres factory made vases of white porcelain painted to resemble veined marble, which were then mounted and were probably more costly than those of real stone. Many of Wedgwood's creamware versions were made with ornamental moulding, swags and handles, gilded in imitation of the metalwork on the French originals. The gilding has proved unable to withstand the wear-and-tear of two centuries, and the unadorned cream-coloured pottery is now revealed, so that the rich appearance of the vases when new can only be imagined.

Although their manufacture was initiated before Wedgwood moved into his new premises, most surviving examples of the creamware vases were made there. The earlier specimens were decorated with painting on the surface and may, generally speaking, be distinguished from those of later date. These were made of semi-blended coloured clays, and were veined throughout the material in the manner of the old Staffordshire agate ware.

Once again Wedgwood was troubled about his London representation, this time with the emphasis on selling. In a letter to his partner, dated 'Sunday morning' and probably written in late May 1767, the case was stated for a change of style and of premises:

I find I did not sufficiently explain to you my reasons for wanting a *Large* room. It was not to shew, or have a large stock of ware in Town, but to enable me to shew various Table and dessert services completely set out on two ranges of Tables, six or eight at least such services are absolutely necessary to be shewn in order to *do the needful* with the Ladys in the neatest, genteelest & best method. The same, or indeed a much greater variety of setts of Vases sho^d. decorate the Walls, & both these articles may, every few days, be so alter'd, revers'd, & transform'd as to render the whole a new scene, even to the same Company, every time they bring their friends to visit us. I need not tell you the many good effects this may produce, when business, & amusement can be made to go hand in hand. Every new show, Exhibition, or rarity soon goes stale in London, & is no longer regarded after the first sight, unless utility or some such variety as I have hinted at above continue to recommend it to their notice.

Clients, he continued, would cease to call and bring their friends if the display remained unaltered:

This may be avoided by us with very little address, when we have a Room proper for the purpose. I have done somthing of the sort since I came to Town & find the immediate good Effects of it. The first two days after the alteration we sold

QUEEN's WARE and ORNAMENTAL
VASES, manufactured by Josiah Wedgwood,
Potter to her Majesty, are sold at his Warehouse,
the Queen's Arms, the Corner of Great Newport Street,
Long Acre, where, and at his Works at Burslim in Staf-
fordshire, Orders are executed on the shortest Notice.
As he now sells for ready Money only, he delivers the
Goods safe, and Carriage free to London.
☞ His Manufacture stands the Lamp for Stewing,
&c. without any Danger of breaking, and is sold at no
other Place in Town.

19. Advertisement from a newspaper of 1769.

three complete setts of Vases at 2 & 3 Guineas a sett, besides many pairs of them, which Vases had been in my Rooms 6-8 & some of them 12 months, & wanted nothing but arrangement to sell them.

Within a few days he thought he had found what he wanted in West-minster and again in Pall Mall, the latter having been used for auction-sales, but both were rejected for one reason or another.

Then, in the spring of 1768 he was again in London, and on 24 March wrote to Bentley announcing that he had taken the lease of a house in St Martin's Lane. 'It was', he said, 'beyond my most sanguine expectations,' and he had already engaged a china-painter 'just come out of Yorkshire,' conjectured to be David Rhodes of Leeds, who would in due course be accommodated at the new address. Hardly had this letter been sent when, just a week later, another followed to notify a change of plan: 'I have met with another house which pleases me better than that I have taken.' The second of these was No. 1 Great Newport Street, at the top of St Martin's Lane, and after negotiations had been concluded it was opened to the public in August of that year.

In addition to all his pressing preoccupations in Burslem and London, Josiah was not free from personal anxiety. His brother John, on a visit to London in June 1767, had been to Ranelagh Gardens and after watch-ing the fireworks display called at the Swan Inn at Westminster Bridge. He left there at midnight and nothing further was known of his move-ments until 5 a.m. when his body was recovered from the Thames. Josiah was very affected by this sad event, being as he wrote to a friend 'rather too susceptible to grief'. John was nine years senior to Josiah, and forty-six years of age when he died.

Just a year later, Wedgwood found that his right leg was increasingly giving him pain. Eliza Meteyard wrote that he had made up his mind to have it amputated prior to the opening of the new pottery, the operation being performed on 28 May 1768. William Cox, in London, was informed of the event in a brief sentence from a Burslem employee: 'Mr. Wedgwood has this day had his leg taken of, & is as well as can be expected after such an execution.' The patient recovered quickly, and within a short time was hard at work again.

Vases continued to be an exciting preoccupation, and once the black ones were perfected their manufacture began at the Bell Works. Production could not keep pace with demand: Wedgwood wrote from London in November (1768) to say that 'Cox is as mad as a march Hare for Etruscan Vases, pray get a qu$^{ty}$. made or we shall disgust our good customers by disappointing them in their expectations'. In the same letter he related to Bentley that he had been with Boulton and together they had called at 'Harraches', perhaps a relative of Pierre Harache, the Huguenot silversmith. There, they saw some old vases purchased recently in Paris, 'I bid £30 for 3 p$^r$. of Vases, they asked £32 & wo$^d$. not abate a penny. There's spirit for you! – must not we act in the same way?' It would be no exaggeration to say that Wedgwood was then living in a wonderland of vases and was almost in a state of obsession with them. In his own words: 'I have lately had a vision by night of some new Vases, Tablets &c. with w$^{ch}$. Articles we shall certainly serve the *whole World*.'

20. Basaltes vase and cover with satyr's mask handles and engine-turned body, *c*.1775; impressed 'Wedgwood & Bentley'. Height 18.2 cm. ($7\frac{1}{5}$ in.). Wedgwood Museum, Barlaston.

CHAPTER THREE

# *Etruria – I*

IN 1769 all that related to vase-making was transferred from the Bell Works to the new manufactory, which was opened ceremonially on 13 June. To mark the occasion the partners made six Etruscan vases, with covers and upright handles, Wedgwood performing the throwing while Bentley provided motive power for the wheel. The vases were sent to London for painting with classical figures above the words '*Artes Etruriæ Renascuntur*' ('The Arts of Etruria are revived') with a commemorative inscription on the other side.

In honour of the area of Italy where the much-admired pottery vases were thought to have originated, the estate and factory were named 'Etruria', Wedgwood's new house being Etruria Hall and Bentley's the more prosaic Bank House. The Etruscan objects, although excavated in the part of north Italy known as Etruria were later found to have been imported there from Greece, but this fact was not accepted until after the building and naming of the manufactory had taken place.

The subjects selected for the decoration of the First Day Vases were taken from a book illustrating ancient vases belonging to William (later, Sir William) Hamilton who had assembled more than one outstanding collection of vases. Hamilton served as British envoy to the Court of Naples for thirty-six years, a post that allowed him an unrivalled opportunity to pursue his divergent interests in volcanoes and objects of art. He began having engravings made of his vases in the middle of the 1760s, their issue in four volumes taking place in 1766–7. Proofs of the engravings circulated prior to publication, some of them reaching Wedgwood by way of the ninth Lord Cathcart, whose wife, Jane, was Hamilton's sister.

Another key source was the '*Receuil d'antiquités égyptiennes, étrusques, gréques, romaines, et gauloises*' by the Comte de Caylus, published in six volumes at Paris between 1752 and 1767. Wedgwood was loaned a copy of the work; then he and Bentley purchased one for their own use. From time to time they acquired other books illustrating ancient vases and antiquities, many of which inspired the shapes and decoration of Etruria's productions.

The partnership between Josiah Wedgwood and Thomas Bentley related only to Etruria and the goods intended for manufacture there,

21. Vases from a design by Jacques Stella (1596–1657): (*Left*) in basaltes, *c.* 1775; 'Wedgwood & Bentley' disc mark. Height 30.5 cm. (12 in.). (*Right*) in variegated creamware, *c.* 1775; height 27.6 cm. (10⅞ in.). Castle Museum, Nottingham.

Bentley having no financial interest in the tablewares and other 'useful' articles being produced at the Bell Works. In September 1770 the distinctions to be drawn regarding the output of the two places were still the subject of discussion between the two men, and Wedgwood wrote:

With respect to the difference between *Usefull ware* & *Ornamental* I do not find any inclination in myself to be over nice in drawing the line. You know I never had any idea that the *Ornamental ware* sho[d]. not be of 'some use'.

The writer then referred to a query by Bentley as to whether his friend's other partnership, that with Thomas Wedgwood at the Bell Works, precluded the making of 'Stellas ewers' at Etruria; the ewers in question being copied from a design in the *Livre de vases* by Jacques Stella, a painter and engraver who lived from 1595 to 1657. Several paragraphs of argument followed, with Wedgwood summing-up his ideas in these words:

May not usefull ware be comprehended under this simple definition, of such vessels as are *made use of at meals*. This appears to me the most simple & natural

41

22. Encaustic-decorated basaltes vase made on the occasion of the opening of Etruria on 13 June 1769; height 25.4 cm. ($9\frac{9}{10}$ in.). Wedgwood Museum, Barlaston.

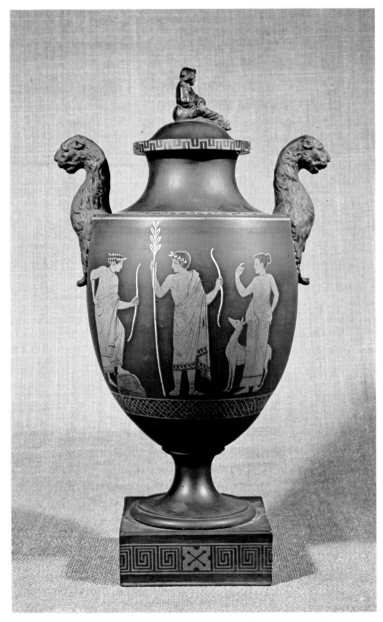

23. Basaltes vase and cover, c. 1775, the encaustic decoration copied from an engraving of a vase in the possession of Sir William Hamilton; height 38.1 cm. (15 in.). Castle Museum, Nottingham.

line & though it does not take in Wash-hand basons & bottles or Ewers, & a few such articles, they are of little consequence & speak plain enough for themselves; nor wo$^d$. this exclude such superb vessels for sideboards, or vases for desserts if they could be introduc'd, as these articles wo$^d$. be rather for *shew* than *use*.

A footnote in Bentley's handwriting on the original letter indicates that thenceforward all was well: 'The Difficulty was easily setled . . .'

The vases excavated in northern Italy were decorated in red on a black ground, or vice versa, by a method that had become forgotten by the 18th century. For imitating them it was necessary to develop some suitable enamels that would emerge from the kiln with little or none of the customary gloss. Wedgwood experimented to produce pigments that remained matt, while he worked also on a glaze to simulate bronze. Both goals were finally attained, and in November 1769 Wedgwood was granted a patent for 'The purpose of ornamenting earthen and porcelaine ware with an encaustic gold bronze, together with a peculiar species of encaustic painting in various colours, in imitation of the antient Etruscan and Roman earthenware'. The use of the term 'encaustic' in the circumstances was confusing, as although the word means literally 'burnt-in' it was, and is, normally applied to the ancient method of painting on a panel with coloured waxes that were melted into the wood.

A Hanley potter, Humphrey Palmer, was soon found to be marketing his own versions of decorated black vases comparable to those of Wedgwood. The latter took out an injunction against the pirate; Bentley's advice was sought as well as that of other friends, and it was realised that the case was less strong than had at first appeared. Although the vases from both sources bore painting copying some of Hamilton's engravings, it would be difficult or impossible to prove that Palmer's had been taken from a Wedgwood vase or direct from the book. In the end an agreement was reached and the action withdrawn, but the occurrence was only one among numerous cases when Palmer and others profited unfairly from Wedgwood's innovations.

Although all arrangements had been made for Bentley to come from Liverpool to live at the house specially built for him at Etruria, he never did so. Instead, in September 1769 a house was leased at Chelsea where he was to live and superintend a group of painters, who would decorate the wares sent down from Staffordshire. The success of his vases was even greater than Wedgwood anticipated, and his letters at the time refer continually to their production and the difficulties to be overcome.

. . . I know you will be very impatient if you have not some plain Vases in London in a fortnight. On rec$^t$. therefore of your l$^r$. on Saturday I sent Moreton immediately to the Wheel for Black Vases though it was near six & made him

44

stay & throw a qu$^{ty}$. We have dry'd them this good Sunday ready for the Lathe tomorrow morning & I hope to have some doz$^{ns}$. fired this week.

This letter was penned in September 1769 and at the end of the year the same subject was recurring:

We find our large Ovens very inconvenient for Vases, I mean in point of time as it is near two months work to fill the bisket oven. I am therefore building a small one to hold two or three basketfull, say £100 Or so of Vases . . . . We have some Medallion Vases in the Oven, & are making plenty of them which you shall have in due time . . .

For a while there were worries over finding a sufficient number of competent and reliable painters for employment at Chelsea. The porcelain factories furnished a few, including Ralph and Catherine Wilcox. Mrs Wilcox was the daughter of Thomas Frye, one-time manager of the Bow Porcelain factory, but she was not anxious to return to the capital

24. (*Left*) Basaltes vase with encaustic decoration, the vase made at Etruria on 13 June 1769 and painted at Chelsea; height 25.4 cm. ($9\frac{9}{10}$ in.). The reverse side is shown on page 42. Wedgwood Museum, Barlaston.

25. (*Right*) Marbled creamware vase, *c*. 1770; in relief on a disc 'WEDGWOOD & BENTLEY: ETRURIA'. Height 20.3 cm. (8 in.). Wedgwood Museum, Barlaston.

26. Encaustic-painted basaltes vase of Egyptian Canopic form, c. 1770; height 37.7 cm. (12½ in.). Traditionally a gift from Josiah Wedgwood to Dr Erasmus Darwin, but possibly given by Josiah II to his brother-in-law Dr Robert Waring Darwin. Wedgwood Museum, Barlaston.

after having worked in the country at Worcester. 'I must talk to her again', wrote Wedgwood on 20 September 1769, and as a consequence of his persuasion the Wilcoxes went to Chelsea. Bentley was informed on the same occasion, 'I hired an ingenious Boy last night for Etruria as a Modeler'. The latter was William Hackwood, who stayed with Josiah and his successors until 1832.

Although the Etruscan vases were then so prominent, if only because they were something entirely new, creamware (alias crystalline, variegated or pebble) continued in production. Early in 1770 Wedgwood defined them to Bentley as follows:

Pebble vases. Suppose we call those barely sprinkled with blue and ornaments gilt, *granite*; when veined with black, *veined granite*; with gold, *lapis lazuli*; with colours and veined, *variegated pebble*; those with colours, and veined without any blue sprinkling, *Egyptian pebble*.

On another occasion he referred to 'Holy Door' and 'Jaune Antique'. Miss Meteyard wrote that the first-named was supposed to be a rich mixture of light puce or mauve varied by gold and white, and the second a rich saffron and black. She attempted to clarify the situation with a brief list:

| | |
|---|---|
| Serpentine | Grey and green |
| Agate | Brown and yellow, with sometimes grey and white |
| Verde antique | Dark green, grey and black |
| Green jasper | Green and grey |
| Grey granite | White and black |
| Red porphry | White on red |

Wedgwood was still busy with his tablewares, ensuring that they were up to his standards and that orders were filled promptly. He was anxious to increase the sale of the whole range of goods, remarking to Bentley in August 1770: 'We are looking over the English Peerage to find out *lines ⅋ connections* – will you look over the Irish Peerage with the same view – I need not tell you how much will depend upon a *proper ⅋ noble* introduction.' He knew that if royalty and the nobility approved of his wares, then everyone else would demand them. In the same year building began at Etruria so that the making of useful wares could be transferred from the Bell Works, and all activities would be under the one roof.

Many of the vases, whether creamware or Etruscan, were mounted on bases of black or white pottery or, occasionally, of marble. Each article was affixed to its base by means of an iron or brass threaded post with a flat top and a nut to hold all firm. Miss Meteyard mentioned that the metal parts were sometimes sent down to the capital from Etruria, and that 'a man named Palethorpe also effected much of this work in

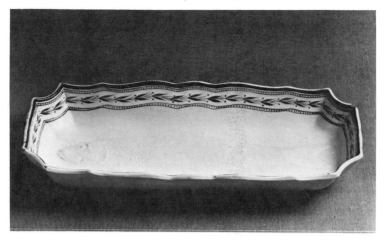

27. Creamware herring-dish, moulded interior and painted border, *c*. 1780; impressed 'WEDGWOOD'. Length 28 cm. (11 in.). Wedgwood Museum, Barlaston.

London, and charged from 6d. to 1s. 6d. for fixing vases and plinths together . . .'

The painting-shop at Chelsea did not confine its activities to vases and ornamental pieces, but also decorated tablewares. Jewitt owned and printed a list of payments made to painters employed there in October 1770, the two cash columns being headed 'On J. W.'s Acct.' and 'On W. & B.'s Acct', thus making it clear that pieces from the Bell Works were treated separately from those of Etruria.

A series of simple patterns evolved over the years for the ornamentation of services, plates and other items being given borders of Etruscan derivation with repeats of such motifs as anthemions or husks. Alternatively there were flower, fruit and leaf designs that are no less reflective of the taste of the period and of Josiah himself. Crests with or without full coats of arms were added to order, only occasionally being placed centrally, as was so often the case with imported Chinese porcelain. Most of Wedgwood's armorial and other wares would have been unlikely to have offended the late R. L. Hobson, of the British Museum, who wrote in *The Connoisseur* in 1908; 'one feels that crests and shields and mottoes, or, in fact, any pictorial decoration, ought to be removed from the gravy and relegated to the rim'. It says much for the creamware that plates, dishes and other tablewares could be produced in their tens of thousands, and that they could be sold with a minimum of coloured decoration; their appeal lay basically in the excellent shaping that was both functional and attractive, and in the unblemished glaze.

In 1768 Lord Cathcart was appointed Ambassador-Extraordinary to

28, 29. Painted jugs ordered by Colonel Legh in 1786; impressed 'WEDGWOOD'. Heights 18.2 cm ($7\frac{1}{10}$in.) and 19.5 cm. ($7\frac{4}{5}$in.). Wedgwood Museum, Barlaston.

the Empress Catherine of Russia, Catherine the Great to some and 'the devil at Petersburgh' to Horace Walpole and others. Wedgwood received orders to supply dinner and dessert services bearing the Cathcart crest: in heraldic terms, 'a dexter hand couped above the wrist and erect proper, grasping a crescent argent'. He informed Bentley on 24 March: 'I have spent several hours with L$^d$. Cathcart our Embassador to Russia, & we are to do great things for each other.'

As a result of this parley, the Empress and many of her wealthy subjects made extensive purchases of tablewares and other goods, and in late 1773 Catherine began to negotiate through her Consul in London, Alexander Baxter, for a very large and elaborately decorated table service. The whole was to comprise nearly 1,000 pieces, each to be decorated with a view in England and to bear the crest of a frog; the latter denoting that it was for use at the Chesmenski Palace, which was in an area known as La Grenouillière (once called the Frog-marsh), near Leningrad.

Wedgwood was delighted to have received such an important commission, but cautiously discussed with his partner the enormous amount of work involved and its cost. He was also anxious about the fate of the order if a war or some political upheaval should occur suddenly in Russia before the account had been settled. In the end it was decided that the potential advertising value of the service was so great that the risk was worth taking.

The painting of the many pieces and the selection of the subjects were the basis of much correspondence between Staffordshire and London.

30. Queensware jelly mould and cover, *c.* 1790; impressed 'WEDGWOOD'. Greatest height 24.1 cm. (9½ in.). Jelly was poured into the plain outer mould, the decorated part being inverted in it; when set, the mould was removed so that the painting was seen through a coating of jelly. Wedgwood Museum, Barlaston.

A few samples were decorated with the principal scene in full colour, but it was quickly found that the cost would be prohibitive, and a colour described variously as 'mulberry', 'delicate black' and 'purple' was adopted for the views. The frog, painted a bright green, was placed in the border at the top. The partners wondered whether each view should be purposely drawn, but Wedgwood saw the impossibility of this, asking Bentley in a letter of 29 March 1773:

Do you think the subjects must be all from *real views & real Buildings*, & that it is expected from us to send draftsmen all over the Kingdom to take these views – if so, what time, or what money? wo^d. be sufficient to perform the one, or pay for the other.

In July (1773) it was decided that Wedgwood should have sketches made of suitable buildings and views in the vicinity of Etruria, and he told Bentley that he had had the offer of a loan of 'Wilsons views from different places in Wales', adding:

There is another source for us besides the *publish'd views*, & the real Parks & Gardens. I mean the paintings in most Noblemens & Gentlemens houses of real Views, which will be sketch'd from by some of our hands at less expence than we

31. Creamware dish with the crest of Leightonhouse in the border, *c.* 1790; width 29.2 cm. (11½ in.).

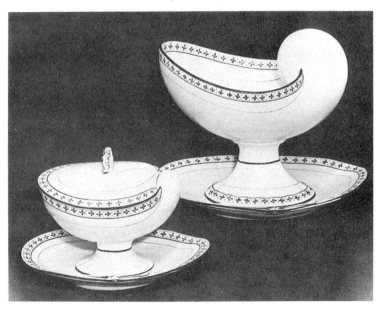

32. Creamware centrepiece and stand and tureen, *c.* 1790, cover and stand of 'Nautilus' shape with painted border; impressed 'WEDGWOOD'. Width of large stand 29.8 cm. (11¾ in.). Wedgwood Museum, Barlaston.

can take real Views, but I hope prints may be picked up to go a great way or we shall be sadly off . . .

Towards the end of the year came mention in a letter from Etruria of a topographical artist named Stringer, of Knutsford, Cheshire:

He is still with me, & I suppose we shall continue taking views for ten days or a fortn'. longer, it being about two days work to fix upon a situation, take a rough sketch, & copy & finish another from that, which is the course he takes.

However, the principal source was engravings, of which there was no shortage and that were easy for the decorators to copy. Wedgwood was careful to include views of the seats and properties of his patrons and friends: 'L$^d$. Gower, & some others of my Neighbours, will expect to be shewn upon some of the larger dishes, & their views are taken accordingly.' No doubt this had been Stringer's task.

The great service was completed in the middle of 1774, and before its dispatch to Russia was placed on display to the public at the London showroom. Among those who saw it there and recorded the fact was Mrs Delany, who rarely missed visiting or commenting on a fashionable event during her long lifetime. She mentioned the occasion in a letter to her niece, Mrs Port of Ilam, Derbyshire, dated 7 June:

33. Plate from the Empress Catherine of Russia's service, *c.* 1774, painted with a view of Mount Edgcumbe, Plymouth, adapted from an engraving published in 1756; impressed 'WEDGWOOD'. Diameter 24.1 cm. (9½ in.).

I am just returned from viewing the Wedgewood-ware [sic] that is to be sent to the Empress of Russia. It consists I believe of as many pieces as there are days in the year, if not hours. They are displayed at a house in Greek-street, called 'Portland House'; there are three rooms below and two above filled with it, laid out on tables, every thing that can be wanted to serve a dinner; the ground the common ware pale brimstone, the drawings in purple, the borders a wreath of leaves, the middle of each piece a particular view of all the remarkable places in the King's dominions neatly executed.

She complained that 'A view of Thorp Cloud, belonging to Mr Adderley' was incorrectly titled, as she was familiar with the hill at the entrance to Dovedale, Derbyshire, knew it was the property of her niece's husband, and informed the person in charge of the room of the error.

The premises in Greek Street, known as Portland House because the ground landlords were the Dukes of Portland, were leased by the firm to replace both the showroom in Great Newport Street and the painting-shop in Chelsea. Portland House was opened with the exhibition of the Russian service which lasted for two months from 1 June 1774, admission being by tickets that were advertised as being obtainable free on application to the former showroom. In July, Bentley moved to 11 Greek Street, adjoining Portland House, where he was well placed to give his attention to all departments of the London end of the partnership, which were now under one roof.

Wedgwood wrote to his partner early in July saying he was glad to learn that the service was being prepared for packing. He then turned to the business of payment, hoping Bentley would be careful not to omit any of the numerous expenses incurred that were unforeseen when the initial estimate was compiled, adding somewhat apprehensively:

I hope one other trespass upon your patience and good nature will complete the whole. I mean settling the price of this work with our good patron, Mr. Baxter – you are much better qualified to do this than myself in many respects . . .

In due course the service reached its destination and the bill for it was paid. The cost to the Empress is unknown, but has been calculated to have been approximately £2,700, to which the actual unpainted pottery contributed a mere £51 8s. 4d. Although the makers may have made only a small immediate profit in money terms, unquestionably they derived long-lasting benefit from the publicity surrounding the commission. Interest was aroused not only by the service itself, its appearance and magnitude, but by the notoriety of its buyer. Exhibiting the entire service in London at the height of the Season was a good stroke of business, paralleling the display in 1763 of the costly Chelsea porcelain service ordered by Queen Charlotte and King George III as a gift to the Queen's brother.

34. Basaltes portrait plaques, *c.* 1780, of (*left and right*) Cornelis and Jan de Witt, Dutch statesmen, and (*centre*) the American politician and philosopher, Benjamin Franklin. The first two impressed 'WEDGWOOD', the last 'Wedgwood & Bentley'. Height 10 cm. (4 in.). Victoria and Albert Museum, London.

35. Pair of basaltes cassolettes (vases forming candlesticks), one with the lid reversed, *c.* 1790; impressed 'WEDGWOOD'. Height 28 cm. (11 in.).

In 1779, Sir James Harris, knighted in 1778 and later first Earl of Malmesbury, British Ambassador at St Petersburg 1777 to 1782, noted in his diary how favourably he was treated by the Empress. On one occasion in 1779 she took him on a visit to La Grenouillière and showed him the Wedgwood service. It then vanished from record until 1909, when Dr G. C. Williamson devoted a book to its description and illustration. At the time of publication, Tsar Nicholas II permitted a number of pieces to come to London to be shown publicly at premises in Conduit Street. Afterwards these specimens were returned to Russia, but it has been suggested that a few of them somehow failed to do so, or that they 'escaped' during or soon after the Revolution. There survive also in the West a few of the trials painted in full colour, and others decorated in monochrome, but lacking the frog crest. A portion of the original service is now housed in the Hermitage at Leningrad, where it is admired by visitors no less than when it was first seen in London.

The year before the Russian service was finished saw the publication of a sixty-page catalogue of Wedgwood and Bentley's productions. The title page announced its contents in general terms:

A Catalogue of Cameos, Intaglios, Medals and Bas Reliefs, with a general account of Vases and other ornaments after the antique; made by Wedgwood and Bentley, and sold at their rooms in Great Newport Street, London.

The introduction included descriptions of the different types of pottery resulting from Josiah's ceaseless trials, many of them made in the few years following the opening of Etruria in 1769.

To give an idea of the *nature* and *variety* of the productions of our ornamental works, it will be necessary to point out and describe the various *compositions* of which the forms, &c., are made, and to distinguish and arrange the several productions in suitable *classes*.

The *compositions*, or bodies, of which the ornamental pieces are made, may be divided into the following branches:—

I. A composition of *terra-cotta*, resembling porphyry, lapis lazuli, jasper, and other beautiful stones, of the vitrescent or crystalline class.

II. A fine *black porcelain*, having nearly the same properties as the *basaltes*, resisting the attacks of acids, being a touchstone to copper, silver, and gold, and equal in hardness to agate or porphyry.

III. A fine white biscuit ware, or *terra-cotta*, polished and unpolished.

The term 'terra-cotta' was used as literally translated, meaning baked earth, and the vases in Class I were the creamware ones described earlier. The 'black porcelain' of Class II was the hard pottery used for making Etruscan vases and other articles, which were described henceforward as basaltes'. The 'white biscuit' of Class III was the white counterpart to the preceding and was also left unglazed; it and the basaltes could be polished by a lapidary without difficulty.

36. Unglazed (biscuit) caneware shell-shaped tureen, cover and stand, c. 1800s. Impressed 'WEDGWOOD'; width 15.5 cm. (6⅛ in.). City Museum & Art Gallery, Stoke-on-Trent.

Two other compositions produced from c. 1775 onwards were caneware, which was buff-coloured and was finished with and without a glaze; and rosso antico, which was a terra cotta red and was more often than not glazeless. These and others when finished in the biscuit or unglazed state were known as 'Dry' bodies.

Since early in the 1770s a fresh field had been increasingly occupying Josiah's attention: cameos or gems, and ornamental tablets and medallions. The tablets were, as stated in the 1773 catalogue, for insetting in chimneypieces and furniture, and in the first-named instance would supplant plaques of carved marble; they would be no less decorative and considerably cheaper. The use of pottery in place of carved wood in connexion with furniture was a fresh departure, but judging by the small number of surviving examples did not become very popular. Alison Kelly in *Decorative Wedgwood* illustrates some pieces of the years 1775–80, among them a *bonheur du jour* or lady's writing-table, heavily inlaid with cameos that are almost certainly of basaltes with painted backgrounds and painted relief figures. More successful in design is a set of eight armchairs once in the Library at Appuldurcombe Park, Isle of Wight, which are of Greek-inspired pattern, each of the top-rails being centred in a small Etruscan portrait medallion depicting a writer.

Ancient carved stone cameos, or gems as they were termed alternatively, had for a long time been eagerly sought by connoisseurs. In the early 17th century the second Earl of Arundel, in the forefront of English art-collectors of his day, gathered together not only a famous gallery of marbles, some of which are now in the Ashmolean Museum, Oxford, but a no less notable series of gems. These duly found their way into the collection of the third Duke of Marlborough, whose descendants sold them and others, known collectively as 'The Marlborough Gems', in 1875.

Wedgwood began by making his cameos from the basaltes with which he had so successfully made his Etruscan vases; he had first tried them in 1769 as ornaments on the latter, terming them 'Medallion Vases'. The 1773 catalogue listed no fewer than 285 different cameos and their counterparts, intaglios, which had their ornament incised instead of in relief and were therefore usable as seals.

As early as 1771 Wedgwood mentioned some of these small objects. He had been persuaded by Mrs Frances Crewe, who lived in Cheshire, to allow Hackwood to model a head of her son, which he reported to Bentley in a letter of 7 September, adding: 'I mentioned Bracelets rings and seals to her, with which she seemed much delighted, & to these I think we may add Gemms to be set in snuff box tops . . .'

Within two years the sale of small cameos for mounting as jewellery had grown into a profitable side of the business. The seals, which could be used for sealing letters and were often carried on the person in the way of a dress accessory, were usually of basaltes polished on the wheel. The cameos were sometimes of the same material, or else of the white biscuit listed in the catalogue.

Designs for the intaglios and cameos came from many sources. At first, the majority originated in those produced by James Tassie, a Scottish gem-engraver who began life as a stone-mason. Tassie employed an opaque white glassy 'enamel' for his reliefs, which were sometimes completed either with backgrounds of the same material or were mounted on ordinary glass backed with tinted paper. Wedgwood bought moulds of modern portraits and ancient gems from Tassie, but he also made his own moulds from original gems belonging to collectors like the Duke of Marlborough. Quantities of the pottery ones were sent to London and Birmingham to be mounted in metal; a letter of 18 May 1773 bears the postscript 'We have sent about 50 doz. of seals to Birmingham this week . . .' No doubt cameos were being turned into jewellery at a comparable rate.

The range of subjects of the portrait cameos widened with the years, and by 1787, when the sixth edition of the catalogue was published, they included a large number of personages of all periods. Among them were heroes of Egyptian, Greek and Roman mythology; philosophers, poets and orators of those cultures; Caesars with their empresses,

37. Metal-framed basaltes plaques of King George III and Queen Charlotte, modelled by William Hackwood, *c.* 1775. Signed on the truncation 'W H', and impressed 'Wedgwood & Bentley', *c.* 1780. Height 8 cm. ($3\frac{1}{8}$ in.).

38. Basaltes portrait plaque of William Shakespeare, *c.* 1780, modelled by William Hackwood in 1777; impressed 'Wedgwood & Bentley'. Height 12.7 cm. (5 in.). Wedgwood Museum, Barlaston.

popes, kings of England; and 'Illustrious Moderns', who included royalties of many lands, statesmen and commanders, philosophers, naturalists, poets, painters and most of Wedgwood's patrons. They could be purchased in sets, singly, or complete with cabinets designed and made to hold them. Each item, cameo or intaglio, was sold enclosed in a paper printed with the name of the subject represented and the catalogue number. When Eliza Meteyard wrote her *Handbook* in 1875, she mentioned that she knew of two collectors who owned cameos still retaining their original wrappings.

Both the basaltes and the white biscuit not only took a polish on the wheel but could be given painted and fired decoration. Matthew Boulton thought of a quicker and cheaper way of colouring the white ones, proposing to tint them with ordinary watercolours and mount them under glass for use as buttons. Wedgwood did not approve of this and wrote agitatedly in December 1773 to his partner:

I do not know how to manage with Mr. Boulton abot grounding cameos in Watercolours, for he is one of those high spirits who likes to do things his own way, & if crossed in that will perhaps do nothing at all, or worse. I will attempt to confine the water colouring to cameo buttons, but if he will order plain white cameos it will be dangerous to refuse him . . .

Josiah suggested there was a possibility that if he refused to let Boulton have the goods he wanted, then he might turn to Worcester for a supply and a good client could be lost. In the end, when Josiah was able to make his cameos with permanently coloured grounds, such imitations were soon routed.

Although always occupied with his experiments, Wedgwood did not neglect the staple profit-earner, creamware. As early as 1767–8 he had the idea of preparing a quantity of samples to send to dealers and agents, and this was continued for many years. The samples took the form of wooden boxes fitted with internal divisions to hold seven or more plates of different sizes and shapes with varying border patterns. Each box was accompanied by a detailed price-list and a set of engravings showing the various articles currently obtainable. It was then a simple matter to order single items or complete services, and the list stated that special designs could be executed if requested. Wedgwood's eagerness to promote business, even in unlikely places, led him to read the published *Letters* of that venturesome traveller Lady Mary Wortley Montagu. He wanted to learn how the Turks lived, and in a letter of 19 September 1772 commented on the matter to Bentley in a less serious vein than usual:

. . . between the Windows in the Ladies Harams are little Arches to set *pots of Perfume*, or *baskets of flowers* Alias Beaupots – Pray are they *Beau* pots or *Bough* pots? – now as there are double rows of windows in these rooms, & Arches between every window, what is a single chimney piece in our solitary rooms to twenty or thirty of these charming little Arches, if they were but comatable for us – Let who will take the Sultanas if I could get at these delightfull little nitches, & furnish them . . .

The matter of pricing goods came to the fore in April 1773, when the local potters held a meeting and resolved to reduce their prices by twenty per cent, making the best table plates only two shillings a dozen wholesale. Did Bentley think the firm could hold their own when charging five shillings apiece retail, while everyone else might be selling at 2s. 6d. or 3s.? Wedgwood put the argument to his partner:

We must endeavour to make our goods better if possible – other people will be going worse, & thereby our distinction will be more evident but I should be glad to have your opinion on our having plates of two prices at the Warehouse [showroom], Viz, at 3/6 & 5/. . . . or shall we drop the best to 4/– or continue as we are?

At Etruria a further use was found for basaltes in the making of portrait busts. When carved in marble or cast in plaster they were popular in libraries, where it was not uncommon to find likenesses of ancient and contemporary authors in the same rooms as their books. Wedgwood saw that his pottery was considerably cheaper than marble and more durable than plaster, and that basaltes with a slightly polished surface was as attractive as either.

As with the cameos, a start was made by buying ready-made examples,

39. Creamware condiment set and stand with relief ornament, c. 1780; impressed 'WEDGWOOD'. Height 23 cm. (9 in.).

40. Three basaltes busts: *(left)* Francis Bacon, from a plaster model supplied by Hoskins & Grant, *c.* 1780, impressed 'Wedgwood & Bentley'; *(centre)* Dr John Fothergill, *c.* 1790, modelled by John Flaxman in 1781; impressed 'WEDGWOOD'; *(right)* Hugo Grotius, *c.* 1780, impressed 'Wedgwood & Bentley'. Heights 48.2 cm. (19 in.), 43.8 cm. (17¼ in.) and 50.8 cm. (20 in.).

in this instance made of plaster. James Hoskins, who was moulder and caster in plaster to the Royal Academy, and his partner Samuel Euclid Oliver, supplied Cicero and Horace in 1770, and by 1773 twenty others had been added to them. More were acquired after 1775, when Hoskins had obtained a fresh partner, Benjamin Grant, the subjects being mostly 17th- and 18th-century English writers. Wedgwood referred to them when writing to London in August 1774:

We are going on very fast now with the Busts having four of our principal hands allmost constantly employ'd upon them. You will find our Busts much finer & better finish'd than the Plaister ones we take them from – Hackwood bestows a week upon each head restoring it to what we suppose it was when it came out of the hands of the Statuary.

Characteristically, he added: 'Pray do not let our labour be unobserv'd when they are under your care.'

While the range of Etruria's 'ornamental' goods was constantly increased, the creamwares from the Bell Works were not forgotten and fresh products were tried if they appeared likely to be profitable. On 20 March 1775 Josiah Wedgwood wrote from London to his cousin to discuss supplying some cisterns for water closets:

... the sale of them will in all probability be very considerable, and the purchasers will be willing to pay a good price for them, from four to six guineas apiece. They now give nine or ten guineas for Marble Cisterns & do not order these of our ware for the cheapest, but because they will answer sweeter than marble ones . . .

The order included a wooden model of the size of the finished article and a set of pegs of the requisite diameter. Wedgwood carefully explained the procedure to be followed in making the cisterns:

. . . I think the best method of making your mould will be to Chop and make their model rough outside and then coat it with Plaster, making the Coat as thick as you think will be sufficient to allow for shrinkings, that is if you think it will shrink four inches then you will coat it two inches thick, . . . and you must like wise allow for shrinking in the size of the holes, for the pegs sent with the model are the real size of the Cocks which are made for the Cisterns.

The amount of shrinkage during firing had to be allowed for in both pottery and porcelain, varying according to the composition of the body. There would seem to be no further record of the cisterns and it can be assumed they were successfully manufactured, but none has survived.

As long ago as 1771 a showroom had been opened in Dublin, and was managed by a man named Brock for a number of years. Eliza Meteyard stated that good business was done there at the start, but this fell off and the enterprise then caused continual anxiety. In her words: 'The resident gentry were, as a body, too encumbered by debt, and too much mixed

41. Basaltes figure of William Hogarth's dog 'Trump', *c.* 1790.
Length 28 cm. (11 in.).

up with the social troubles and party strife of their country, to have money to spend on luxuries . . .'

The year after the opening of the Dublin premises, Bentley went to Bath to seek a suitable outlet there. He settled on some rooms in Westgate Buildings, and on hearing the news his partner wrote:

I am glad you have taken Rooms. I should know what time they are intended to be open'd as it will require six weeks or two months to prepare a sortment of usefull ware here, services, &c, & to get them to Bath by water.

In June 1772 Wedgwood took his wife, who had been seriously ill, to Bath in the hope that her health would be improved by the waters and the change of air. While he was there he wrote to London saying that he had called at 'a very rich shop in the Market place', where there was a fine display of Boulton's goods. He requested Bentley to send down a further supply for their showroom as it was essential for the firm to exhibit a complete selection of their ware, 'or we had better do nothing for I think the Toy, & China shops are richer and more extravagant in their shew here than in London.'

Sarah Wedgwood made a slow recovery, but Josiah suffered anxieties not only on account of his wife's illness and his own disability. His employees and their grievances contributed their share. The men started to complain about the piece-rates they were receiving, which Wedgwood had decided before he left for Bath were too high in view of the changed state of trade. Sales of vases showed a severe drop, and, as he put it in a letter of 23 August 1772:

The Great People have had their Vases in their Palaces long enough for them to be seen and admir'd, by the *Middling Class* of People, which Class we know are vastly, I had almost said infinitely, superior in number, to the Great, & though a *great price* was I believe at first necessary to make the Vases esteemed *Ornaments for Palaces* that reason no longer exists. Their character is established, & the middling People would probably buy quantitys of them at a reduced price.

Wedgwood pointed out that overheads ran on unabated and the only way in which he could reduce his expenses and lower the price of his goods was by 'making the greatest quantity possible in a given time'. Bentley would have understood the argument, but it was far from easy to convince the workers that if they produced more at a lower rate per item their overall wages would not be reduced.

There was trouble, too, in London, where the head clerk, Benjamin Mather, was reported by an anonymous correspondent to have been embezzling the firm's money. Wedgwood had been suspicious of him because his accounts were always behindhand, and for some months he did not appear to have drawn his wages although he must have had money from somewhere. Instant dismissal was ruled out: 'he may go and collect thousands from our Creditors'. Relief was to be discerned in a letter written a week later (7 September 1772):

I shall be very glad, as well on the young Man's account as my own, to know that he has not gone any great lengths in either vice or youthful folly, and most sincerely wish he may be reclaimed before he is gone beyond recovery.

The matter continued to be discussed by the partners, Josiah considering that it would be wrong to discharge Mather peremptorily, and it would be preferable to encourage him to reform his ways. He was sent down to Bath so that he might help Mr and Mrs Ward, who managed the showroom, while the stock and accounts at Greek Street were checked, and was later given responsibility of reporting on the firm's Amsterdam agent. There, he found that someone else had unknowingly been following his example, and having sold goods was making no attempt to pay for them. In October 1777 Mather again fell into evil ways and was said to have been spending his time drinking instead of doing his work. Again he was given a stern lecture in the hope of reforming him. Finally, in August 1780, after yet another effort to bring him to his senses, Mather's eleven years of employment, in which normal behaviour too often alternated with periods of disgrace, terminated in dismissal and he is no longer heard of. The partners could not have tried harder or been more patient with someone who so often rejected their compassionate gestures.

# CHAPTER FOUR

# *Etruria – II*

THE 'white biscuit' comprising Class III of the 1773 catalogue was the result of many experiments to perfect imitations of gems. As early as the middle of the 1760s Wedgwood had made trials with some clay from South Carolina sent to him by a friend in Manchester, precisely the same as some received by a local potter who said 'he could make nothing at all of it'. However, Josiah was less condemnatory, remarking that the material would require special treatment and future use of it depended on the quality and price of supplies.

The clay in question had been known in England since at least 1744, when the first patent taken out by Thomas Frye and a partner for making porcelain at Bow included in its ingredients 'an earth, the produce of the Cirokee [sic] nation in America, called by the natives unaker . . .' A further patent four years later did not contain it. In 1767, a couple of years after receiving his first supply, Wedgwood realised that the clay had possibilities, and began attempts to obtain a quantity of it for further experiments. He was very anxious to keep the matter as secret as possible, for fear of piracy by other plotters, and consulted the Duke of Bridgewater and other highly placed persons as to how he should go about achieving his object.

By chance, Wedgwood visited a friend whose brother had been recently in South Carolina, and who offered to return there in the role of agent to the potter. Details were settled and the man, Thomas Griffiths, set sail in the late summer for Charleston, Carolina, some 300 miles from the clay beds which lay in country belonging to the Cherokee Indians. Although Griffiths obtained some clay and sent it back to Staffordshire, it did not serve the required purpose any more than some other American clay obtained from Pensacola, Florida. Wedgwood discussed this with typical scientific attention to detail in a letter to Bentley of 21 November 1768:

It must be got as clean from soil, or any heterogenous matter, as if it was to be eat & put into good casks or boxes, & if they were to get several parcels at different *depths*, & put them in separate casks, properly number'd, I could by this means easily ascertain what depth of mine is best for our purpose, as it is very probable that there is a great difference in that respect, if the stratum be a thick one.

Research continued over the years, and by the middle of 1774 he had

42. Flower vase with relief figures of 'The Seasons' after John Flaxman, of white jasper with traces of gilding, *c.*1780; impressed 'WEDGWOOD'. Height 16.5 cm. (6½ in.). Castle Museum, Nottingham.

got on the track of what promised to be success. Wedgwood discovered that barium sulphate was the ingredient required, and with its aid was able to make a fine white stoneware that was ideal for his needs. The barium was obtained from Derbyshire, where it was known as 'cawk', and he was worried lest his secret should be discovered before he had reaped any benefit from it. Wedgwood stated that he dared not have it sent direct to Etruria, but suggested to Bentley that it might go to London and there be ground to a powder to disguise it prior to forwarding to Staffordshire. So anxious was he that in January 1775 he wrote: '. . . if it was sent by the West Indies the expence would not be worth naming in comparison with the other considerations'.

The new composition had a particular advantage in that it could be left white or coloured to choice. A month before the preceding letter Wedgwood had stated to Bentley how progress was being made:

From several late series of experiments, I have no reason to doubt being able to give a fine white composition any tint of a fine blue, from the Lapis Lazuli to the lightest Onyx. I propose making the heads of this composition, and sometimes the Grounds, but each separate. By this means we shall be able to under-cut the heads a little. The Grounds must be cut even and polished, and the under side of the heads ground so as to lie perfectly flatt upon the polish'd ground, and then Mr Rhodes must fix them with a little Borax &c in his enamel Kiln.

43. Jasper vase and cover with a relief of 'The Apotheosis of Homer'
designed by John Flaxman, *c.* 1790; impressed 'WEDGWOOD'. Height 47 cm.
(18½ in.). Castle Museum, Nottingham.

44, 45. Metal-framed blue and white jasper portrait plaques of King George II and Queen Charlotte, *c.* 1775; impressed 'Wedgwood & Bentley'. Height overall 8.6 cm. (3⅜ in.). Wedgwood Museum, Barlaston.

I am now making a few blanks and heads, of which I have not the least doubt but you will make the finest things in Europe.

On New Year's Day 1775 there was more news for Bentley. After saying that he was glad the white body had met with approval in London and that he was confident he had mastered his production, Wedgwood continued:

The blue body I am likewise *absolute* in of almost any shade, & have likewise a beautiful Sea Green, & several other colors, for grounds to Cameo's, Intaglio's &c. . . .

A fortnight later a further difficulty was reported after attempts had been made to fire cameos in once piece, when it was found that the background colour tended to stain the white relief. This defect was duly overcome; the plaques had their subjects affixed with a thin slip and were successfully fired complete.

In November 1775 Wedgwood referred to his new material for the first time by the name that has remained with it, 'Jasper', adding that he intended to make gems from it suitable for mounting as 'Bracelets' Rings, &c.' Not for a further two years was he fully satisfied, writing to Bentley on 3 November 1777.

I have tried my new mixing of Jasper, & find it very good. Indeed I had not much fear of it, but it is a satisfaction to be certain, & I am now *absolute* in this precious article, and can make it with as much facility and certainty as black ware.

68

He did not exaggerate: during the ensuing decades jasper cameos were made in their tens of thousands for a multiplicity of uses for which pottery had never before been employed. By 1779, many hundreds of different cameos and intaglios were in production, and there was a comparably wide range of portrait medallions of ancients and moderns. In addition there were plaques for insetting in snuff boxes and furniture; drums for the bases of candelabra and for opera glasses; buttons, beads and bell pulls; gems for chatelaines and watch keys; and not least in size and importance, tablets for chimneypieces and wall decoration.

Jasper made a shy first appearance in the 1774 catalogue, where it formed class IV, and was described as 'A fine white *terra-cotta*, of great beauty and delicacy, proper for cameos, portraits, and bas reliefs'. In 1787, by when it had become established, the description was much less inhibited:

IV. Jasper – a white porcelain bisqué of exquisite beauty and delicacy, possessing the general properties of the basaltes, together with that of receiving colours through its whole substance, in a manner which no other *body*, ancient or modern, has been known to do. This renders it peculiarly fit for cameos, portraits, and all subjects in bas relief, as the ground may be made of any colour throughout, without paint or enamel, and the raised figures of pure white.

For a period prior to the introduction of jasper there was a com-

46. Sword hilt and guard of cut steel inset with blue and white jasper cameos and beads, *c.* 1790. Castle Museum, Nottingham.

47. Pair of gilt metal and cut glass candlesticks with jasper drum bases,
*c.* 1790; impressed 'WEDGWOOD'. Height 28 cm. (11 in.). Wedgwood Museum,
Barlaston.

48. Jasper vase and cover, 'The Prince of Wales Vase', bearing a relief portrait of the prince, later Prince Regent and then George IV, *c.* 1785–90. Height 37.4 cm. (14¾ in.).

position described in the 1779 edition of the catalogue as 'A white waxen biscuit ware'. Although it possessed many of the virtues of jasper, the latter proved to be of consistently higher quality and so it supplanted the 'waxen biscuit'. As regards colour, the pale cobalt to which Josiah Wedgwood's name is attached to this day, was the most popular, but the palette duly included dark blue, sage-green and brownish olive-green, lilac, yellow, and an intense black that differed from the basaltes.

A few years after the two-colour cameos went into regular production,

49. (*Left*) Cut steel buckle inset with blue and white jasper cameos: (*left*) 'Bourbonnais Shepherd' and (*right*) 'Poor Maria', both after designs by Lady Templetown. Height overall about 8 cm. (3⅛ in.). Castle Museum, Nottingham.

50. (*Right*) Blue and white jasper-dip plaque of Hercules and the Nemean Lion, *c.* 1800; impressed 'WEDGWOOD'. Height 2.6 cm. (1⅛ in.).

continuing losses through discolouration during firing led in part to a further change. The jasper background had hitherto been coloured throughout the body, but the spotting in addition to the high cost of cobalt led Wedgwood to introduce 'Jasper-dip'. For this, a white plaque was given a thin coating of coloured jasper on one or both sides, in the latter case making a sandwich of blue-white-blue that could be polished on the edges if required. Jasper-dip was produced from *c.* 1780, but solid jasper returned to favour and it is not possible to date examples solely from these indications.

The quick success of jasper led to a need to increase the number of subjects for reproduction. In the selection of designers Wedgwood was at pains to employ the best available; artists who were not only proficient at their task, but whose style suited the medium. His most renowned modeller was John Flaxman, whose father, also named John, supplied Etruria with a number of plaster casts of reliefs and busts. The son, born in 1755, gained prizes for drawing when he was only eleven, and at fifteen attended the Royal Academy school. There, came second in the 1772 competition for the Gold Medal, being beaten by Thomas Engleheart who was ten years older and subsequently achieved little. On this occasion the self-assured John Flaxman was very disappointed and showed it, for Wedgwood apparently referred to this weakness when writing to Bentley in January 1775: 'I am glad you have met with a Modeller, & that Flaxman is so valuable an Artist. It is but a few years since he was a most supreme Coxcomb, but a little more experience may have cured him of this foible.' A bill of 1775 printed by Eliza Meteyard refers to the supply of bas reliefs of mythological characters, a set of

'The Seasons', and vases, signed 'for my father, John Flaxman, Jun<sup>r</sup>.'

The younger Flaxman's work for Wedgwood in the years 1775–85 included portraits as well as tablets and a set of chessmen. In 1775 he was working on a likeness of Dr Daniel Solander, the Swedish botanist, and another of Joseph Banks (later, Sir Joseph and President of the Royal Society) to whom Solander was secretary and librarian; both had sailed with Cook in the *Endeavour* and became nationally known. Flaxman also made bas relief portraits of the Duchess of Devonshire; Herschel, the astronomer; Charles James Fox; Captain Cook; William, son of Benjamin Franklin; Sarah Siddons; and John Philip Kemble. The larger compositions included the well-known 'Dancing Hours', 'Hercules in the Garden of the Hesperides', this at a cost of £23, and 'Apollo and the Nine Muses', of which four of the figures were modelled in 1775 and the others two years later.

Among others who supplied material to Wedgwood in the 1770s for portrait plaques and other items were: Joachim Smith, who first achieved a name for himself when he presented George III in 1762 with a wax model of the infant Prince of Wales lying naked on a couch; John Bacon, later to achieve eminence as a sculptor; Theodore and Richard Parker, probably father and son, who rendered accounts for casts of medallions and figures; Mary Landré, who was paid for plaster casts and '4 Boys in metal' at 7s. each; and two men better-known as modellers of porcelain figures; Pierre Stephan, of Derby, and 'Tebo' (Thibault?) of Bow, Plymouth, Bristol and Worcester, neither of whom earned praise from Wedgwood for their work.

While he was so busy conducting the lengthy experiments that led to the triumph of jasper, Wedgwood was by no means free of other worries. One of these was concerned with Richard Champion, manager of the

51. One of a pair of blue and white jasper plaques depicting 'Apollo and the Nine Muses', after John Flaxman, c. 1780; impressed 'WEDGWOOD & BENTLEY'. Width 40 cm. (15¾ in.).

Bristol porcelain manufactory, which had been started in 1770 following the closure of Cookworthy's venture at Plymouth. William Cookworthy, an apothecary, had recognised in Cornwall the China stone that would combine with china clay to make a true hard porcelain of the Chinese or Meissen type, and had been granted a patent for his process in 1768. In 1773, Champion became owner of the Bristol works and two years later, when the seven-year life of the patent was due to expire attempted to renew it for a further fourteen years.

52. Jasper plaque of 'Diomedes gazing at the Palladium', *c.* 1775; impressed 'WEDGWOOD & BENTLEY'. Height 20.3 cm. (8 in.). So long as the Palladium, an image of the goddess Pallas, remained in Troy, the city was safe from capture: Diomedes duly took it. Wedgwood Museum, Barlaston.

The original patent specified that the use of the stone and clay should be limited exclusively to Cookworthy, then by purchase to Champion, and the extension was sought in the House of Commons. Wedgwood organised local support to oppose it; in particular he had the aid of John Turner, owner of a successful pottery at Lane End, a few miles south-east of Newcastle-under-Lyme. The argument against the monopoly was on several grounds, but in the main it was propounded that natural materials should not be restricted to any one user and so inhibit progress. In the end, a compromise allowed Champion use of the clay and stone in the manufacture of porcelain, but the Staffordshire potters were free to employ them for other products. The Bristol concern proved uneconomic and the patent was sold in 1781 to a company of Staffordshire potters who in turn, gained little benefit from it.

The bill relating to Champion's patent received royal assent on 26 May 1775, and on the 29th of that month Wedgwood, accompanied by John Turner and Thomas Griffiths, the latter having returned from America, set out for Cornwall. Josiah kept a journal in which he made notes of what he saw and did during the expedition, the purpose of which was, in his own words:

I thought it would be proper to take a journey into Cornwall, the only part of the kingdom in which they [the stone and clay] are at present found, and examine upon the spot into the circumstances attending them, – whether they were to be

53, 54. Two blue and white jasper portrait plaques: (*left*) Joseph Priestly attributed to William Hackwood, *c.* 1785, impressed 'W E D G W O O D', (*right*) Sir Hans Sloane, *c.* 1785, modelled from a carving in ivory by an amateur, Silvanus Bevan; impressed 'W E D G W O O D'. Height of each 8.9 cm. ($3\frac{1}{2}$ in.). Castle Museum, Nottingham.

55. 'Labourers' painted by George Stubbs in fired colours on a biscuit earthenware plaque, 1781. Dimensions 69.8 × 91.4 cm. (27½ × 36 in.).

had in sufficient quantities, – what hands they were in, – at what prices they might be raised, &c. &c.

In due course the travellers reached Truro, the day before one of the periodical sales of locally mined tin, and ingots of the metal were displayed in the streets. Wedgwood had been warned to be on his guard when he was among the Cornish miners, who had a strong animosity towards Staffordshire potters in general and Josiah in particular, because their wares had displaced pewter at the table and so lowered the price of tin. However, the travellers encountered no difficulties and were able to arrange for supplies of what they sought at prices they were prepared to pay, leaving Thomas Griffiths at St Austell to supervise affairs on their behalf. By 16 June Wedgwood was back at Etruria where, after a brief rest he was again hard at work pursuing his trials with the addition of the newly obtained materials.

Two years later, Wedgwood became associated with the animal painter, George Stubbs, who had approached him for the provision of large plaques for his work. Stubbs had begun to experiment with enamel-painting and sought pottery as a possible base in place of the more usual copper. In December some progress had been made in their manufacture.

We have fired 3 tablets at different times for Mr. Stubs, one of which is perfect, the other two are crack'd & broke all to pieces. We shall send the whole one (22 inches by 17) on Saturday & are preparing some larger.

A year later, trials were still proceeding, Wedgwood telling Bentley: 'I mean to arrogate to myself the honour of being his *canvas maker*'.

By May 1779 it would seem that Stubbs had received a fair number of usable plaques, so Wedgwood wrote to his partner about payment for them. The account was settled in kind, by paintings on pottery, a group portrait of the Wedgwood family in oils on a wood panel, and two models in relief for reproduction in basaltes or jasper: 'Phaeton and the Chariot of the Sun', and 'A Lion preparing to attack a Horse'. The pottery plaques were made of the 'white biscuit' modified to enable them to emerge from the kiln in suitably large sizes. To have produced them was a remarkable achievement, and while it may be argued that Stubbs used an idiosyncratic method of picture-making, the £300,000 realised by an example at Christie's in 1978 suggests that they appeal to modern taste.

George Stubbs stayed at Etruria while painting Josiah and his family and other portraits, and a side-effect of his visit was the production of artists' palettes. Some small-sized palettes were sent to London for sale at one shilling apiece, Wedgwood informing his partner: 'we are making some large ones under Mr. Stub's direction . . . we are now making some doz$^{ns}$. for your rooms'.

Always aware of the vital importance of being one move ahead of the public, Wedgwood was sensing that buyers might be tiring of his Queensware and demand something new. In 1779 he introduced his 'Pearl White' to augment creamware, which had been marketed successfully for over a decade. As he wrote to Bentley in August of that year, he was perfectly satisfied himself with the well-tried favourite, 'but you know what Lady Dartmouth told us, that she and her friends were tired of creamcolor, and so they would of Angels if they were shewn for sale in every chandlers shop through the town'. The new ware was given some of its whiteness by the use of a glaze containing a trace of cobalt; which resulted in a perceptible blueness where the glaze gathered, such as round the foot-rims of plates.

It was at about the same date that Wedgwood was able to perfect a composition for the making of heat-resistant crucibles, and other articles for use by scientists and chemists. His life-long interest in the two subjects was shared with a number of men whom he met from time to time, or with whom he corresponded. He was friendly with, and supported financially, the clever and controversial Joseph Priestley, to whom Wedgwood was probably introduced by the Reverend William Willett, whose wife was Josiah's youngest sister. Willett had a good

56. (*Left*) Blue and white jasper portrait plaque of Thomas Bentley (1730–80), modelled *c.* 1773; impressed 'WEDGWOOD', *c.* 1785. Height 8.9 cm. (3½ in.). Victoria and Albert Museum, London.

57. (*Right*) Black and white jasper medallion lettered in low relief round the border 'Am I not a man and a brother?', modelled by William Hackwood in 1786. Impressed 'WEDGWOOD', *c.* 1800. Height 3 cm. (1¼ in.).

knowledge of science, while Priestley had an even better understanding of the subject and temporarily became a dissenting minister in order to pay his bills. The range of apparatus produced at Etruria was gradually extended to include retorts, syphons, filter funnels and evaporating pans, some of which were made to the designs of his friends. The articles were supplied free of cost for a number of years, but as knowledge of this benevolence grew widespread and demand rapidly increased, Wedgwood was forced to make a charge to all except a small circle that included Priestley.

The correspondence between Wedgwood in Staffordshire and Bentley in London terminated abruptly in 1780, the last letter from the former being dated 12 November. A brief notice among the obituaries in the *Gentleman's Magazine* of that month announces the reason.

26. At Turnham-Green, Mr. Bentley, in partnership with Mr. Wedgwood. For his uncommon ingenuity, his fine taste in the arts, his amiable character in private life, and his ardent zeal for the prosperity of his country, he was justly admired; and will long be most sincerely lamented.

Bentley had moved in 1777 from his house in Greek Street to Turnham Green, where he lived near his friend, Ralph Griffiths, whose brother Thomas had travelled to Cornwall with Wedgwood and Turner, who was founder and owner of the *Monthly Review* and possibly author of the lines quoted above.

There is no record of the cause of Bentley's sudden passing at the age of fifty. He was twice married, his first wife dying in 1759 when giving birth to a child who died soon afterwards, and he married a second time in 1772, his widow surviving him. It is much to be regretted that no more than a very few of Bentley's letters to his partner have come to light. That they were retained and revered by their recipient is clear from a mention that Wedgwood kept them; according to his great-grand-daughter, Lady Farrer, he had them 'bound up in volumes known in the family as "Josiah's Bible", on account of his having them so constantly by him.' Without the letters being available today it is not possible to assess accurately the role of Bentley in the partnership, but it is clear that Wedgwood relied on him as a confidant, and that he played a not unimportant part in the promotion of Etruria's products in England and far afield.

Following the death of Thomas Bentley, the stock of finished goods held jointly by the two men was disposed of by auction 'in concurrence with the wishes of Mrs. Bentley'. The forthcoming dispersal was adver-tised in advance, an announcement in the *Sherborne Mercury* of 19 Novem-ber 1781 concluding with the rider:

N.B. The Table and Dessert Services, and the various other useful things made in Queen's Ware, being the property of Mr. Wedgwood alone, continue to be sold at his Rooms, as usual . . .

The same notice appeared also in the sale catalogue.

The auction was conducted by Messrs Christie & Ansell, predecessors of the present firm of Christie's, at their premises in Pall Mall, the sale beginning on 3 December and continuing for a further eleven days. A total of approximately £2,250 was realised by the 1,200 or so lots: an average of just under £2 apiece, which surely gave satisfaction to no-one except the buyers.

Each lot in the sale comprised a number of articles of similar type, in some instances making a quantity that might seem to be of little interest to an average purchaser. Thus, lot 15 was:

> Twelve *Tea-pots*, three *Cream-ewers*, four *Sugar-dishes*,
> two *Basons*, one *Canister*, one *Butter-tub* and *Stand*

all of which were knocked-down for £1 5s. The buyer was entered as 'S.N.' with the name Green written at the side. A noticeable proportion of the lots went to 'S.N.' and 'N', both penned in a flowing script, lead-ing to a supposition that these were perhaps *noms de vente*, either for Wedgwood himself or for one of Christie's staff acting on behalf of buyers wishing to remain anonymous.

Some of the purchasers were less reticent and can be identified. Among them was George, second Baron Vernon, who gave £1 15s. for a set of five basaltes vases ornamented with festoons, and Miss Fauquier, who became his second wife in 1786, who purchased busts of Demosthenes and Cicero for £2 12s. 6d., suggesting that she and her husband had in common a liking for the same make of pottery. Lady Pembroke, of Whitehall, paid no more than £1 5s. for two dozen cameos suitable for mounting as jewellery; but a few days later five Etruscan vases were bought by Sir Joseph Banks for £10 10s, a fair price under the circumstances. Undoubtedly many of the names are those of long-forgotten traders whom it is now difficult to identify. Was 'Yeldham' M. & T. Yeldham, the Russian merchants of New Broad Street, and was 'Partridge' the Samuel Partridge with a Staffordshire warehouse in St Swithin's Lane?

Bentley certainly agreed with Wedgwood in political and religious matters, and was probably the more advanced thinker of the two on these subjects. After he left Liverpool and his newly founded Octagon Chapel, he remained in touch with its members, the 'Octagonians', and deplored the closing of the building in 1776. His partner supported nonconformity, including among his portrait-medallions likenesses of the Unitarian minister at Newcastle-under-Lyme (his own brother-in-law) and John Wesley, as well as a series of heads of the popes. Wedgwood's outlook on religion was described succinctly by Professor C. H. Herford: 'Religion, in the theological sense, had little part in Wedgwood's life. His religion was to make perfect pots' and, it should be added, to sell them.

In the matter of politics, although Bentley had made his living in Liverpool, which was the centre of the slave trade in England, he was a strong abolitionist. Wedgwood followed in his footsteps by patronising the anti-slavery movement, and went so far as to get Hackwood to model a cameo from the seal of the committee of the Society for the Abolition of the Slave Trade, which was produced in jasper and distributed far and wide. Other causes espoused by the potter included the American struggle for independence, represented by portraits of Franklin and Washington, and the French Revolution, the revolutionaries being depicted on medallions in the act of storming the Bastille and jeering at Louis XVI as he returned to Paris. Wedgwood cultivated the acquaintance of Whig politicians, who did their best to make his products known, assisted him in obtaining material for copying, and gave him advice when required.

To take Bentley's place in London, Wedgwood sent his nephew Thomas Byerley, son of Josiah's widowed sister Margaret. Byerley had been a lively youngster and was an actor for some time before his uncle arranged in 1768 for him to go to Philadelphia. The sum of £70 was to be available for him on arrival there, Wedgwood writing to Bentley asking him to give the lad the benefit of his counsel, adding:

If he has a mind to take a few pieces, or 3 or 4 pounds worth of my Ware with him, you may let him have so much on his own acc^t., & pray advise him to ship himself in the cheapest way for America, as he may there come to want a Crown, or even a shilling, & know not where to have it.

When Byerley returned to England in 1775 after having worked in New York as a teacher, he was given employment at Etruria. As he could read and write French he was of great assistance, and he was sufficiently confident of his Latin to offer to teach the language to Josiah's children. However, in this respect Wedgwood wrote to Bentley in 1780:

I wish you would be so good to inquire of your friends who may be likely to know of a latin Master for Mr. Byerley is too much engaged to be depended upon if he were ever so well qualified for the purpose.

One of Wedgwood's interests in the 1780s was the establishment of the General Chamber of Manufacturers of Great Britain, a body started to promote the interests of the country's manufacturers. Not least, it set out to discourage, if not prevent, the emigration of trained workers to foreign countries, whether on the mainland of Europe or America. This problem had come to Wedgwood's attention when some of his own men had left Etruria for employment overseas, and in 1783 he went

58. Blue and white jasper teaset and tray, c. 1790, (later known as a Cabaret) with reliefs after designs by Lady Templetown; impressed 'WEDGWOOD'. Width of tray 34.6 cm (13⅝ in.). Buten Museum of Wedgwood, Merion, Pennsylvania.

59. The Entrance Hall, Carlton House, London home of George, Prince of Wales, an engraving from W. H. Pyne's three-volume work, *The History of the Royal Residences*, published in 1819. In the spaces above the rows of columns at each side are vases and busts that may have been supplied by Wedgwood.

so far as to write and print a pamphlet addressed to the remaining work-force, warning them of the perils that might overtake them if they joined their former colleagues. The chamber played a part in 1785 in drafting a commercial treaty with the Irish, but this was abandoned following a lukewarm reception by both parliaments. More important to the potters were the treaties with France and Saxony, which resulted in a lowering of the high duties charged hitherto on English pottery sent across the Channel.

A few months prior to the signing of the French treaty in October 1787, Wedgwood sent his son John with Thomas Byerley to Paris, where they arranged for Dominique Daguerre to sell wares from Etruria. Daguerre, of the rue Saint-Honoré, was the foremost dealer in Paris, supplying furniture as well as expensive decorative objects to the Courts and nobility of many countries. He had obtained a near-mono-poly in buying Sèvres porcelain tablets for mounting on furniture, and may have considered doing the same with jasper. However, when the cases of jasper and Queensware arrived from England he seems to have had most of them put in a cellar, made little or no effort to market the

60. Jasper flower vase, *c.* 1790; impressed 'WEDGWOOD'. Width 21.3 cm.
(8¾ in.). Castle Museum, Nottingham.

creamware, which formed the majority of the consignment, and the venture proved a failure. At the time of the Revolution Daguerre came to London, opening premises in Sloane Street where, among other clients, he numbered the Prince of Wales who had earlier patronised his Paris establishment. The Prince also gave his custom to Wedgwood, as among his numerous outstanding debts in 1795 for goods supplied earlier to Carlton House was the sum of £158 owing to 'Wedgwood & Co., of Greek Street'.

The 1780s also saw Josiah Wedgwood honoured by being elected a Fellow of the Royal Society, to which he contributed papers between 1782 and 1790. Some of these were on the subject of the pyrometer that he had devised for measuring the heat inside a furnace. Hitherto, there had been no satisfactory method of knowing the temperature of a kiln, which was well above anything capable of being recorded by a mercury thermometer, and reliance had to be placed on the experience of the fireman tending it.

A welcome newcomer at Etruria was Alexander Chisholm, who had served as assistant to Dr William Lewis, chemist to the Society of Arts. Chisholm became a most useful employee, helping conduct the endless experiments, performing general secretarial duties for Josiah, and

61. (*Left*) Blue and white jasper bowl, *c.* 1785, ornamented with subjects designed by Lady Templetown and modelled by William Hackwood; impressed 'WEDGWOOD'. Diameter 17.8 cm. (7 in.). Wedgwood Museum, Barlaston.

62. (*Right*) Blue and white jasper portrait plaque of Samuel Johnson, *c.* 1800, modelled by John Flaxman in 1784; impressed 'WEDGWOOD'. Height 10.8 cm. ($4\frac{1}{4}$ in.).

helping to educate the children. As the years passed he increasingly acted as private secretary, many of Wedgwood's later letters and note-books being in Chisholm's handwriting.

The range of goods was further widened by the making of vases and other ornamental pieces in jasper ornamented with bas reliefs. Among them were vases and covers for decoration, vases for flowers, candle-sticks, and teawares of such elegant design that they could only have been intended for display in cabinets. A modeller, Henry Webber, son of a Swiss sculptor, Abraham Wäber, was taken on the staff in 1785, although two years earlier, when he won a silver palette at the Society of Arts, he had given his address as Etruria. Webber proved an excellent craftsman and was referred to by his employer as having been recom-mended to him by Sir Joshua Reynolds 'as the most promising pupil in the Royal Academy'. Another Academy pupil, John Lochee, executed portrait-medallions from *c.* 1774.

A few well-connected amateur designers were able to have their work accepted, notably Lady Templetown, Lady Diana Beauclerk and Miss Crewe, the first two specialising in sentimental studies of children at work or play. Lady Templetown's husband was a member of the house-hold of the Dowager Princess of Wales, mother of George III; Lady Diana was the eldest daughter of the third Duke of Marlborough; and Miss Crewe lived not far from Etruria, Hackwood's first essay having

63. Blue and white jasper plaque, c. 1790, formerly called 'The Sacrifice of Iphigenia', modelled in Rome by Camillo Pacetti in 1787 from a relief on the Barberini sarcophagus, in which it was alleged the Portland Vase was found; impressed 'WEDGWOOD'. Width 38 cm. (15 in.).

been to model a likeness of her brother who duly succeeded to the title of Baron granted to his father in 1806.

A further series of bills from John Flaxman reprinted by Eliza Meteyard list some of the work on which he was engaged between 1781 and 1787. Figures, busts, a tureen, portraits, plaques commemorating the French treaty, and a set of chessmen show how varied were his commissions. The portraits included one of Dr Samuel Johnson, whose death occurred on 13 December 1784. That Johnson did not actually sit for him is apparently proved by Flaxman's bill dated 3 February of that year:

A [model in wax] of Dr. Johnson      £2. 2. 0d.
A print of the D$^r$. for assistance in the model      2. 6d.

The seventy-five-year-old Doctor was unwell at the time, writing to a friend on that very day: 'I am still confined to the house, this is the eighth week of my incarceration'. The engraving used by Flaxman is thought to have been one made by Thomas Trotter from his own drawing, and published in 1782.

In the autumn of 1787 Flaxman set out for Rome, Wedgwood advancing him money for the purpose, the artist's objective being to forward his studies while providing drawings and models for the factory. He was joined there for a time by Henry Webber, and also secured the services of a French sculptor, John Devaere, who later came to work at Etruria. Several Italian artists were also recruited by Flaxman, so there was an ample supply of fresh material flowing back to Staffordshire.

64. Jasper chess set, *c.* 1795, designed by John Flaxman in the mid-1780s. Height of king 10 cm. (4 in.). City Museum & Art Gallery, Stoke-on-Trent.

Also received at Etruria from a distance was a consignment of clay from Australia. This was sent in 1789 by Governor Philip on the instructions of Sir Joseph Banks; with it came a mineral substance that Wedgwood analysed and discussed in the journal of the Royal Society. The clay he pronounced to be excellent, some of it being used for making a plaque, designed by Webber and depicting 'Hope addressing Peace, Labour and Plenty', as a compliment and encouragement to the newly settled colonists.

Some years earlier, in 1778 Wedgwood had sent as a gift to Sir William Hamilton at Naples a jasper tablet of 'The Apotheosis of Homer'. He wrote to Bentley at the time: '. . . it might be agreeable to Sir Wm. to shew his friends in Italy the use which had been made of his collection of vases in England & not a bad mode of shewing our productions there . . .' He added that the plaque should be sent in a silk-lined mahogany box; 'I am almost superstitious in the effect of such accompaniments.' In reply, Sir William expressed his thanks and took the opportunity of recommending further subjects that might be similarly adapted.

A few years later Hamilton became closely concerned in the events surrounding the article that was to be the finest and most famed of Wedgwood's productions, and one that provided a triumphant climax

65. Jasper cup and saucer, *c.* 1790; impressed 'WEDGWOOD'. Diameter of saucer 12.4 cm. (4⅞ in.). Probably made for display in a cabinet rather than for use. Castle Museum, Nottingham.

to his career as a potter: the object eventually known universally as the 'Portland Vase', The original vase was made in the late first century A.D., of dark blue glass overlaid with opaque white glass, possibly in Alexandria but generally referred to as of 'Roman' origin. According to tradition it was excavated from a tomb in Rome in the year 1582, and received its first mention in print sixty years later when it was described as being one of the treasures of the Barberini palace in that city.

The vase was further described with illustrations from 1697 onwards, and in the early 1780s, when the then owner of the palace sold it, allegedly because she had contracted large debts from gambling, most connoisseurs and students were aware of its existence. The buyer was a Scot, James Byres, who acted as a guide and banker to visiting Britons and supplied many of them with works of art with which to return home. He sold the vase to Sir William Hamilton who, in 1783 brought it to England and sold it to the Duchess of Portland, a notable collector of all kinds of curiosities. When the Duchess died eighteen months after making the purchase, her collection was sold by auction and the vase bought by her son, the third Duke of Portland.

Wedgwood had known of the vase from engravings and had been eager to reproduce it in jasper, so he must have been particularly elated when he signed a brief document on 10 June 1786. It read, in part:

66. The Portland Vase, *c.* 1790, Josiah Wedgwood's careful copy of the Roman glass original now in the British Museum. Height 25.4 cm. (10 in.). Wedgwood Museum, Barlaston.

I do hereby acknowledge to have borrowed and received from His Grace the Duke of Portland the vase described in the 4155th lot of the Catalogue of the Portland Museum . . . and I do hereby promise to deliver back the said Vase . . . in safety into the hand of his Grace upon demand.

Detailed drawings of the vase had been completed within a twelve-month, Henry Webber being principally concerned with them and with the modelling. It was only after three further long years of patient trial and error that it became possible for Wedgwood to arrange the public showing of an example at Greek Street, at the aptly named Portland House, during the months of April and May 1790. The vase was accompanied by a document declaring it to be 'a correct and faithful imitation, both in regard to the general effect, and the most minute detail of the parts'. It bore the signature of the President of the Royal Academy, Sir Joshua Reynolds, who was probably responsible for the Duke having purchased the original vase at his mother's sale.

As soon as it could be spared from Greek Street the vase was placed in the hands of Wedgwood's second son, also named Josiah, and Thomas Byerley, who set off with it on a tour of Europe. They visited The Hague, Amsterdam, Hanover, Berlin and Frankfurt where they showed the vase, and other Etruria productions, to heads of state, the nobility and anyone else of importance who might be interested. Towards the end of the journey Wedgwood wrote to his son:

I do not know whether I have told you in so many words that you must not on any account part with the vase, but bring it back with you. It will be necessary to keep this identical one that I may be able to confront gainsayers with it. We have not yet made one so fine as yours.

It is thought that about thirty of the vases were made successfully during Josiah's lifetime, but of these some remain untraced. Among the latter is the example taken round Europe, which has been said erroneously to be in the British Museum. In subsequent years the firm made many thousands of copies of the Portland Vase (in the late 1830s with the figures demurely draped), but the 'first edition' is probably the finest, and certainly the most celebrated, of Etruria's wares.

In 1790 Josiah reached the age of sixty, and on 16 January of that year he made his three sons, John, Josiah and Tom, and Thomas Byerley partners in the business. The firm was styled Josiah Wedgwood, Sons and Byerley but at the time John and Tom had little or no desire to become potters. As a result, in June 1793 the firm became for a time Wedgwood & Son & Byerley. During these years old Josiah was in semi-retirement, taking holidays at Buxton, Blackpool and the lakes, and making a tour in Wales. An idea of his scale of living at Etruria Hall is provided in a list made in May 1794: there were seven male servants from butler down to gardener. In addition ten horses were kept two four-wheeled carriages and one two-wheeled.

On 3 January 1795, in his sixty-fifth year, Josiah died. He was buried in the porchway of the parish church of St Peter, Stoke-on-Trent, and when the church was rebuilt in 1829 his remains were removed to the churchyard. A monument to him was placed in the chancel in 1803, and this was duly re-sited in a similar position in the new building; the monument taking the form of a shaped slab of black marble on which is an inscribed tablet of white marble supporting a Portland vase and a basaltes ewer. Above the tablet, carrying the inscription composed by Sir James Mackintosh who combined the callings of philosopher and judge (in the vice-admiralty court at Bombay) and was brother-in-law of John and Josiah II, is a head of Josiah Wedgwood by John Flaxman.

Wedgwood left a considerable sum of money, including £30,000 to each of his sons, £25,000 to each of his daughters, and the Etruria estate with the Hall and manufactory to Josiah. It was a remarkable termination to the career of a man whom Jewitt informed his readers started in life with 'the poor and miserably small sum of twenty pounds', conveniently forgetting that his wife brought him the money with which Etruria was built, and that he doubtless received help, if not also legacies, from his numerous relatives. Samuel Smiles, author of the moralising classic *Self Help*, similarly made the potter the hero of one of his 'rags to riches' books written to encourage late-Victorian youth. Be that as it may, it is beyond argument that, in the formal words on his monument, Josiah Wedgwood 'converted a rude and inconsiderable manufactory into an elegant art and an important part of national commerce' – and so it remains.

67. Enamelled gold mourning ring lettered 'JOSIAH WEDGWOOD'. Diameter 2 cm. ($\frac{13}{16}$ in.). In his will, Wedgwood directed that rings should be given to persons he had named on a list, but the list could not be found. Wedgwood Museum, Barlaston.

# CHAPTER FIVE

# *From 1795*

WHEN JOSIAH died the partners remaining in the business were his son Josiah II and Thomas Byerley, the latter in charge at Greek Street. There, the twenty-one-year lease was on the point of expiry and a move was made to another location, this time to Westminster, to a house at the east corner of St James's Square and the present Duke of York Street. The entrance to the large building was at the side, in what was then called York Street, and a view of the principal showroom, fully stocked and with a salesman welcoming a lady visitor, appeared in Ackermann's *Repository of Arts* in 1809.

Of old Josiah's three sons, two had married the sisters Elizabeth and Louisa Jane Allen of Cresselly, Pembrokeshire, and Tom, the youngest, remained a bachelor. None showed more than a fraction of the deep interest their father had had in Etruria, and as all of them were wealthy they had no compelling reason to do so. When he split from his brothers in 1793, John became a partner in the newly established London bank of Alexander Davison & Co., then of Stratford Place and from 1804 in Pall Mall. When the bank failed in 1816 it was merged with Coutts and bankruptcy was avoided, but John was unable to recover the capital he had invested. He was a restless man, living successively at Cote House, Bristol; Seabridge and Betley, both near Newcastle-under-Lyme; in Herefordshire; and after 1825 in Gloucestershire.

Josiah had shaken the clay of Staffordshire from his feet when he married, moving to Stoke House, Cobham, Surrey, and then, in 1800 leasing from his brother Tom the large estate of Tarrant Gunville, about six miles north of Bridport, Dorset. Finally, in 1807 he returned to his native heaths by going to live for an interval at Etruria and then at Maer Hall, a house situated to the south of Newcastle-under-Lyme and only seven miles from Etruria.

The third brother, Thomas, who had a poor constitution, spent much of his time in scientific research, of which one of the results was an early attempt at photography. He went so far as to succeed in making photographic images, but failed to discover how to render them permanent. Like his elder brother, he found it difficult to settle and was continually on the move, often in vain attempts to improve his health. He bought Tarrant Gunville and almost at once let it to his brother Josiah, then

68. 'Wedgwood & Byerley, York Street, St. James's Square', coloured
engraving from Ackermann's *Repository of Arts*, February 1809.

bought the adjoining estate at Eastbury. Tom and Josiah provided
financial support for Samuel Taylor Coleridge, the philosopher and
poet, by making him an annual allowance of £150. It enabled Coleridge
to study in Germany where, as he wrote later: '. . . instead of troubling
others with my crude notions and juvenile compositions, I was thenceforward better employed in attempting to store my own head with the
wisdom of others.' Thomas died at Eastbury in 1805 at the age of thirty-
four, leaving in his will an annuity of £75 to Coleridge.

In addition to the fact that Etruria must have missed the experienced
leadership of old Josiah Wedgwood, prevailing trading conditions were
suffering because of the Napoleonic war. In 1800 John rejoined the firm,
remaining a partner for twelve years, but giving it only occasional if
keen attention. Then, he found to his alarm that much had deteriorated
over the years; Byerley had done his best to keep things going, but he
was not a practical potter and lacked authority.

Nevertheless, there were a few innovations, among them the so-
called 'pastry-ware': unglazed caneware tureens resembling pies that
would hold cooked meat and could be brought to the table. They had
been suggested as early as 1786, but nothing seems to have come of the
idea for a decade. References were made to these simulations at later
dates by more than one writer, stating that their popularity stemmed
from a shortage of wheat-flour that occurred, according to the reliability
of the authors' memories, in 1795–6 or in 1800.

Copper, silver and gold lustres were also marketed. They had been
known of at Etruria before 1795, but were only perfected and used there
at later dates. They were principally employed on creamware; the

'Moonlight' lustre, comprising gold and various colours applied in the form of marbling, being much admired. On pearlware, a pink lustre was used with good effect.

In order to increase production and keep the manufactory up to date, the creamware department was enlarged and a steam engine installed. This was made and erected by Boulton and Watt during 1801 at a cost of £1,095, improvements being made to it seven years later. The use of the engine was remarked on by Harry and Lucy, Maria Edgeworth's fictional knowledge-thirsty brother and sister, who paid a visit to Etruria in 1825. In the book *Harry and Lucy concluded*, Lucy explains to their ever-attentive parent that 'there was a sort of large roller, in the shape of a cone, mamma, and a strap or band round it, that could slide or be pushed up or down this cone . . . to make the [potter's] wheel go slower or quicker'. In 1844, Alexander Brongniart of the Sèvres porcelain factory, who was at Etruria in 1836, commented on the engine. He mentioned that it manipulated and pumped the clay and that, *'par des communications très-ingénieuses'*, it also served the throwers at their wheels.

By 1804 John had moved from Bristol to Seabridge, from where he could exercise closer supervision. In February 1804 he wrote to Josiah, in Dorset, reporting on some of the confusion he had found:

The whole system of the slip kilns and clay beating has gone to ruin, and nothing will restore it but vigorous measures and these cannot so well be taken as when we are all on the spot. . . . When I was at the works this last week I found the clay was in miserable order . . . The whole system of the ovens also requires a fresh arrangement . . . I am afraid I bore you with repeating these things to you; but they are so strongly impressed on my mind . . .

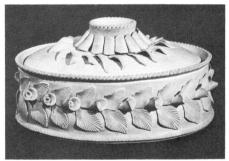

69. Caneware game dish and cover simulating pastry, *c.* 1810; impressed 'WEDGWOOD'. Length 30.5 cm. (12 in.). Buten Museum of Wedgwood, Merion, Pennsylvania.

70. Jardinière and stand of lilac, green and white jasper in a basket-weave pattern, *c.* 1800; impressed 'WEDGWOOD'. Height 8.9 cm. (3½ in.). Buten Museum of Wedgwood, Merion, Pennsylvania.

71. Blue and white jasper group on pedestal, *c.* 1810, the group showing
'Britannia' with foot on prostrate 'France', the pedestal with portraits of
British Admirals, modelled by Henry Webber to commemorate victories
over the French. Height overall 81.3 cm. (32 in.). The whole is seen in the
centre foreground of the 1809 view of the York Street showroom
illustrated on page 92. *Group:* Private collection.
*Pedestal:* Wedgwood Museum, Barlaston.

A year later the same writer gave his brother further news, little of it
offering much encouragement:

About a fortnight ago we had upwards of 60 doz. of plates out of one [glost:
glaze] oven spoiled by the fires, the next oven had 40 doz., and the men seemed
inclined to consider that as a very fair sample of firing; but I declared at once
that I would no longer submit to such work, but would discharge them at once.
This has had its proper effect.

The same letter told how the creamware 'has lost its softness and even-
ness of surface, which constitutes its great beauty', because it was being

72. Soup plate transfer-printed in blue under the glaze with the 'Bamboo' pattern introduced in 1805. Impressed 'WEDGWOOD', *c.* 1805. Diameter 25.1 cm. (9⅞ in.).

fired in the same kiln as pearlware. The latter required a higher temperature than the cream, so he proposed the obvious solution of firing the white on its own. A couple of months later, in April 1805, John was less anxious:

> ... our ware comes out of the Biscuit Oven much better than what it did; the plates more sortable, and the ware in general not scorched as it was before. The Glos[t] men now do their duty and send the ware in very good order, and the firemen become much more regular and steady.

At the end of the year the firm began marketing tablewares printed with patterns in underglaze blue, a type of pottery that many of their neighbours and competitors had been making and selling in increasing quantity for a number of years. The process employed for the decoration was similar to that of Sadler and Green, but in this instance the transfer was applied to the biscuit ware, which was then glazed and fired for a second time. The blue-printed pottery was made in direct competition with the 'Nankin' imported from China, so the earlier patterns were of Oriental inspiration; typical of them was the 'Willow-pattern', introduced by Thomas Minton in the early 1790s and subsequently widely imitated. Etruria entered the market with a design referred to at the time as 'Bamboo', which has been identified as one showing a luxuriant Chinese garden having in the foreground a large vase decorated with a musician – or is he inside the vase *(illus. 72)*.

Then came a series of floral patterns, some in blue but others in brown, most likely due to John Wedgwood's great interest in botany and horticulture. This interest had led to the founding of what is now the Royal Horticultural Society, which sprang from a preliminary meeting held in London in 1804. On that occasion, John took the chair, and following further meetings he was elected treasurer.

One of the brown-printed patterns was the 'Water Lily', often stated to have a connexion with old Josiah's friend, Dr Erasmus Darwin, who died in 1802. Una des Fontaines has shown that this is erroneous, a dinner-service of the pattern, on its first introduction in 1808, being sold in the normal way of business to Dr Darwin's son, Robert. He was married to Susannah, John Wedgwood's sister, and one of their children was Charles Darwin, the eminent naturalist. The service was decorated in underglaze brown, the first use of this colour in printing, embellished with touches of orange and gilding to give a rich effect.

In addition to entering the market with their new blue-printed ware, Etruria continued to produce the old favourites: basaltes, jasper, creamware and so forth. Byerley, then in London, was urging John and Josiah to send him greater quantities of articles and to introduce some new ones. In 1807 he wrote plaintively:

I am sorry to tell you that we appear here to stand in need of all the aid you can give us with Novelties, for, though the town is very full, our rooms are very empty most days. I wish I had something to go to the Queen with. She graciously expects us at this time of the year; but I know of nothing new and good enough.

Later in the same year he advises Etruria: that 'candlesticks for reading, toilettes and washstands' would sell well, particularly if they were in blue and white jasper 'and come cheap'.

In the summer of 1807 Byerley crossed to Dublin, and bought a large house in the city which he converted into a showroom, and which it was hoped might attract the people, or their sons and daughters, who had patronised Josiah Wedgwood nearly forty years earlier. Etruria was informed of how the new rooms were to be set out with a different type of ware in each, 'but I should feel more secure if we had something

73. Pearlware bulb pot with removable top, decorated with printed patterns in three colours and with hand-colouring, made to commemorate the fiftieth anniversary of the accession of George III in October 1809; impressed 'WEDGWOOD'. Diameter 15.2 cm. (6 in.). This once had a loop handle of which one of the fixing-points is seen at front centre. Wedgwood Museum, Barlaston.

*peculiar*, and in this respect the Jasper may still befriend us, for there is very little of it in the market'. Byerley added, 'what we have of it is, I fear, however, more calculated to stand on our shelves and be admired than to produce profit.'

Despite the recurrent complaint that nothing new was coming from Etruria, there was at least one attempt to meet a likely demand. The government declared that 25 October 1809 should be a public holiday, Jubilee Day, George III having been proclaimed King on 25 October 1760 and entering into the fiftieth year of his reign. To commemorate the event, the firm made some suitably inscribed articles, of which a bulb pot is illustrated. The decoration is printed in no fewer than three colours: orange-red, purple and brown, with the oriental flowers hand-painted in colour. Some portions of teasets are known that bear similar decoration; decoration that was completely untypical of Wedgwood, but was a brave essay in the prevailing taste.

Thomas Byerley died at the end of 1810, aged sixty-two. He had spent over thirty years with the Wedgwoods, father and sons, and without doubt could not have helped comparing the exciting times of the elder Josiah with what ensued after 1795. Eliza Meteyard wrote that through no fault of his own he came to bear the major portion of the burden of the entire manufactory and its showrooms for the last fifteen years of his life: he was 'really earnest in his desire that Etruria should not lose her old fame for works of taste; but his natural place was the counting-house, not the studio or the workshop'.

In 1812 came a completely new introduction, something never

74. Tureen and cover printed in underglaze blue with the 'Corinthian Temples' pattern, *c.*1815; impressed 'WEDGWOOD'. Length 36.8 cm. (14½ in.).

75. Bone-china teacup, coffee cup and saucer, *c.*1815. Printed 'WEDGWOOD' in red. Diameter of saucer 14.3 cm. (5⅝ in.). City Museum & Art Gallery, Stoke-on-Trent.

attempted before by a Wedgwood: porcelain. The formula used was similar to that employed by other English makers of the time; a quantity of calcined bone being added to China clay, china stone and flints to produce the material known as 'Bone-china', that remains to this day an esteemed British product. Thomas Byerley's son, named Josiah after his great-uncle, and who had taken over the managership of York Street, wrote to say that he foresaw no difficulty in selling china teasets if they were sent down to him. However, for one reason or another, they did not reach London until June, which was too late in the season. As Wolf Mankowitz has indicated, old Josiah would have launched the new product very differently, and the half-hearted way in which it was introduced to the public shows 'very clearly the limitations of the factory in the years which followed his death'.

Wedgwood bone-china continued to be sold until as late as 1829, but production of it may have ceased at an earlier date with orders being filled from stock. It would seem to have been most in demand in about 1815–16, when the Marchioness of Landsdowne ordered some bowls decorated with a pattern of 'hawthorn buds and leaves, varied by red and green lines,' and a member of the Rothschild family purchased a dinner service decorated with 'blue and gold diamonds, surrounded by blue triangles, edges of the articles in gold'. Both painting and printing were used to decorate china with oriental, floral and other subjects, but

76. Caneware covered pot-pourri jar decorated in enamels with Chinese flowers, c. 1820; impressed 'WEDGWOOD'. Height 24.8 cm. (9¾ in.). Wedgwood Museum, Barlaston.

it would seem that Wedgwood's traditional restrained styles were unable to compete with their rivals' florid patterns.

In 1819 experiments took place for the making of a 'stone china', which was marketed in the year following. It was a dense and opaque ware, described by the late Tom Lyth of Barlaston, as being similar in appearance to French hard-paste porcelain but greyish in tone. Each piece was given a printed mark, together with a pattern number which allays any doubts as to what it is. Decoration invariably took the form of printed outlines filled-in with hand-painted colours and gilding. Manufacture ceased by 1861.

For one reason and another trade had declined so much that it was decided in 1828 to dispose of the London showroom, selling both the premises and the stock. From the summer of that year a clearance sale took place; not only the ornamental and table wares, but such related items as moulds and trial pieces that had accumulated there during the thirty or so years that the firm had been in St James's. The property was sold at the end of 1829 and the firm retreated to Staffordshire without the benefit and prestige of a London outlet of its own.

In the years following 1812, when John Wedgwood relinquished his interest in Etruria, changes continued to occur. In 1823 Josiah II's eldest son, Josiah III, became a partner but failed to take an active role in the pottery. He bought an estate in Surrey, Leith Hill Place, in 1837, retired altogether from the firm five years later and died at the age of eighty-five in 1880. Josiah's third son, Francis, was made a partner in 1827 and settled to the arduous task of reactivating the concern. In the year after the death of Josiah II, which occurred in 1843, the entire Etruria estate, the pottery, village, Hall and all else, was offered for sale by auction, but the manufactory did not change hands. Francis bought himself a one-hundred-acre estate at Barlaston, a few miles to the south of Stoke-on-Trent, where he resided from 1848. In the middle of 1843 he took into partnership John Boyle, who died only eighteen months later, and was followed from 1846 to 1859 by Robert Brown. Three of Francis's sons successively became partners, and members of later generations of the family followed suit, so that today the name of Wedgwood is still to be found among the directors of the company.

The mass-production resulting from the Industrial Revolution and its use of steam-power inevitably led to a lowering of the quality of design and finish. The reaction to this began to appear during the early 1840s, when the long-somnolent Society of Arts, Manufactures and Commerce (later, the Royal Society of Arts) initiated a series of annual exhibitions. Prizes were awarded for the best items in each class of goods, but they did not attract much interest at first. In the third of them, held in 1846, one of the winning entries was a teaset made by Mintons to the

77. White biscuit half-jug, a traveller's sample, *c.* 1820. Height 12 cm (4¾ in.).
Wedgwood Museum, Barlaston.

design of 'Felix Summerly', the pseudonym of Henry (later, Sir Henry)
Cole. He considered there was a ready market for well-designed, good
quality articles, and in the year following launched a firm named Felix
Summerly's Art Manufactures. Cole, manager of the concern, gathered
a number of interested designers, modellers and manufacturers, the
latter to make and market the articles while paying a royalty to Summerly's. In return, the firm commissioned and approved the prototypes
and attended to publicity.

Among the participating firms was the Wedgwood company, of
whose contributions to the venture some can be described. One was a
salt-cellar in the form of a dolphin supporting a shell on its back, made of
pearlware, coloured and gilt, the marks including a printed one of Felix
Summerly's initials with the name of the designer, John Bell, in full. The
other Summerly piece is a set comprising a shaving-pot, brush-dish and
brush-handle depicting 'Heroes Bearded and Beardless', designed by
Richard Redgrave, and shown at the 1848 Society of Arts exhibition.
Redgrave is known also to have designed a porter cup for the Wedg-
wood/Summerly venture.

The Art Manufactures lapsed when Henry Cole became deeply·
involved in the Great Exhibition of 1851 and in this, too, Wedgwood's

played a part. The exhibition was divided into sections of classes according to the goods displayed; class 25 comprising 'China, Porcelain, Earthenware, &c.', shown in the North Transept Gallery of the great iron and glass 'crystal palace'. The firm's display embraced all their current productions, and from the number and variety of the articles in the catalogue it is clear that they made a big effort to regain some of their lost importance. As the entry details what was being manufactured in 1851, and for many years afterwards, it is reprinted here in full:

Carrara (statuary porcelain). – Figures from the antique – Venus and Cupid, 27 inches high; Cupid, 24 inches high; Hercules, 20 inches by 17; Morpheus, 24 inches long; Venus, 19 inches high; Mercury, 17 inches high; Faun with flute, 17 inches high.

Figures – The Preacher on the Mount; crouching Venus; Nymph at the fountain; Cupid and Psyche, group; Cupid with bow.

Triton candlesticks, right and left (Flaxman).

Busts of Washington; Shakspeare on pedestal; and Venus. Sleeping Boys.

Spill cases, 'Bonfire,' and set of three 'Muses'.

Black vase, 8 inches (with the pedestal 9 inches) 'Apotheosis of a Poet'. Two vases, 16 inches each, 'Water and wine'. Various other vases, plain, of different sizes. Lamp and candlestick.

Copy of Portland, or Barberini vase, 10 inches high, dark blue, as the original in the British Museum.

Another in black, with white jasper bas-reliefs. Blue jasper with white bas-reliefs from the antique: – Vase, 25 inches high, with pedestal 10 inches high, 'Sacrifice to Cupid'.

Vase, $27\frac{1}{2}$ inches high, 'Apotheosis of a Poet'.

Vase, with pedestal, $19\frac{1}{2}$ inches, 'Ulysses discovers Achilles'. Another to match, 'Infancy of Achilles'.

Vases and pedestals, 21 inches, 'Muses'. Vases, 12 inches, 'Hunting and Hawking,' and 'The Arts and Sciences'.

Various other vases of different sizes, plain and ornamented, including 'Hercules at the garden of the Hesperides'. 'Muses'. Bacchanalian subjects, rivers, and arabesque designs.

Flower and incense vases and covers. Alumette club-shape pint jug; temple lucifer box; toy watering can; ring, cigar, and pen trays. Ornamented and Venetian lucifer boxes. Violet baskets; round covered, tooth powder, and lip-salve boxes.

Snuff-boxes. Shaving box, with zodiac ornaments.

Toy garden-pots and stands. Set of chess-men; thirty-four pieces, by Flaxman.

Jugs, various, Florence shape with bas-reliefs.

Octagon and oblong smelling bottles, silver mounted.

Howard and Clarendon tea-pots, pint.

Tea-pot, with aquatic plants, pint. Coffee-pot, Amoy shape, quart.

Pillar-shaped candlestick. Piano candlestick. Taper candlestick.

Jasper. – Cameos of various colours, with white bas-reliefs from the antique.

Black, with red Etruscan figures: – Vases, various, and with inscriptions.

78. Jasper vase and cover, *c.* 1850: impressed 'WEDGWOOD'. Height 42.3 cm.
(16⅝ in.). Shown at the Great Exhibition held in London in 1851. City
Museum & Art Gallery, Stoke-on-Trent.

79. (*Top*) 'Round Etruscan' tureen, cover and stand, an engraving in the *1880 Catalogue of Shapes*.

80. (*Left*) Cambridge ale jug and stand, of red pottery with coloured inlaid bands, patented 11 November 1850. Impressed 'Wedgwood' and with date marks for February 1863. Height 16.2 cm. (6⅜ in.). Perhaps named in memory of the Duke of Cambridge, son of George III, who died in July 1850. Buten Museum of Wedgwood, Merion, Pennsylvania.

81. (*Right*) Jug decorated with coloured glazes and inscribed 'What though my cates be poor / Take them in good part / Better cheer may you have / But not with better heart', adapted from Shakespeare's *Comedy of Errors* (III, i, 28–9); impressed 'WEDGWOOD'. Height 24.1 cm. (9½ in.). Possibly made for sale in 1864 when the tercentenary of the poet's birth was celebrated. City Museum & Art Gallery, Stoke-on-Trent.

Red terra-cotta, with black bas-reliefs, from the antique: – Choice vases, varied in style, size and ornament. 'Clarendon' toy tea set; comprising tea-pot, sugar, cream, slop bowl, and bread-and-butter plate. Tea-pot, pint, with Egyptian ornaments.

Red porous earthenware: – Wine and butter coolers, various designs. Water bottles and stands. Jug with cover and stand, quart. Butter cooler, buff porous earthenware.

Chemical earthen and stone ware: – Mortar and pestle, 5 inches; mortar, 1 inch. Evaporating pan, acid proof. White pill tile, graduated. Funnel, fluted; coarse crucible and cover. Voltaic stone-ware trough, with red porous earthenware lining. Porous cylinder, and flat porous cell for voltaic apparatus. Mercury and water baths. Digester. Conical filterer; triangular filter holder.

Plumbers' earthenware: – Closet pans, cream-coloured, and flowing blue printed. Square wash-table, cream coloured. Long square wash-table, with fittings complete, marbled. Wash-basin, with plug-hole and waste pipe, marbled.

Cream-coloured, or Queen's ware, with enamelled borders: – Etruscan-shaped soup-tureen and stand. Round covered vegetable dish. Dinner plates, in various designs.

Cream-coloured earthenware (Queen's ware): – Plates and dishes. High oval soup-tureen and stand, (by Flaxman). Round and oval soup-tureens. Round Etruscan soup-tureen and stand. Round covered vegetable dish. Herring dish, with embossed fishes. Oval twig pattern fruit basket and stand. Oval quatrefoil-pattern fruit basket and stand. Fruit dishes, various shapes. Quart jugs, Dutch and Roman shapes. Bowls, water-ewers, nursery-lamp. Coffee-biggin, with stove and lamp cup. Milk-boiler and cover. Wine-funnel, with strainer. Egg beater; blanc-mange moulds; pudding-cups. Egg-shaped pudding-boiler. Round and oval milk pans. Pierced milk-skimmer. White stone tea-pots, arabesque and wheatsheaf patterns.

Coloured earthenware: – Tea-pot, pint, Rockingham-coloured, tall and low. Tea-cup and saucer, Bute shape, drab colour. Breakfast bowl and saucer, French shape, drab colour. Oval game pie, cane colour, ornamented. Cambridge ale jug, pint red-coloured earthenware. Embossed leafage dessert plates and dishes, green glazed. Twig ornamented fruit basket and stand. Two-handled vase, red enamelled Chinese flowers. Jug, half-pint, club shape, black enamelled Chinese flowers. Small plain red garden pots and stands.

The jury judging class 25, under the chairmanship of the Duke of Argyll, reported that the articles shown by the firm 'are of great as well as long-acknowledged merit – which consists chiefly in a faithful revival of forms originated by the elder Wedgwood, some of the most remarkable of which were suggested by the genius of Flaxman'. The report added that 'no better desire could be entertained for the popular taste of the country in respect to this class of article than it should again be familiarized with these productions . . .' Despite an absence of any mention of new patterns or types of ware, which weighed with the jury in other cases, Wedgwood's were awarded a Prize Medal.

82. Jasper vase of *c.* 1880 with relief medallion of a warrior with the winged 'Victory'; impressed 'WEDGWOOD'. Height 24.1 cm. (9½ in.). Wedgwood Museum, Barlaston.

At the head of the list of the firm's exhibits was their new 'Carrara' ware, which had been produced from 1848 to meet the public demand for white marble-like statuettes. Some years earlier, in 1842, a new variety of biscuit china was put on the market, Copeland's and Mintons each claiming priority for its invention. It was eminently suitable for the reproduction of statuary, and because of its resemblance to the marble quarried on the island of Paros, in the Aegean Sea, it was named 'Parian'. The Wedgwood company, thinking on similar lines, named their composition after the equally famous quarries at Carrara, in Tuscany. In later years, more figures, groups and busts were added to the number shown in 1851.

Eighteenth-century Wedgwood basaltes and jasper were already being sought by collectors, and a selection of specimens was loaned to the important Art Treasures Exhibition, held at Manchester in 1857. Five

83. Majolica plate with pierced border. Impressed 'WEDGWOOD' and date marks for April 1864. Diameter 22.6 cm. ($8\frac{7}{10}$ in.). Wedgwood Museum, Barlaston.

84. Garden stool of pottery with coloured glazes. Impressed 'WEDGWOOD' and with date marks for 1875. Height 46 cm. (18⅜ in.). Temple Newsam House, Leeds.

years later, at a comparable showing of national treasures, this time at the South Kensington Museum, London, a much larger selection of old Wedgwood was displayed. The owners on this occasion included the Chancellor of the Exchequer of the day, William Ewart Gladstone; Isaac Falcke, who presented his collection to the British Museum in 1909; and Dudley Coutts Marjoribanks, later Lord Tweedmouth, whose collection, which included a number of the original wax models for jasperware, was purchased by W. H. Lever, later Viscount Leverhulme, and is now in the Lady Lever Art Gallery, Port Sunlight, Cheshire.

Interest in the old undoubtedly stimulates interest in the new: in this instance, in modern copies of period Wedgwood. However, the firm could not live completely in the past, and had to do its best to compete with its contemporaries. One of these was Mintons of Stoke-on-Trent, who displayed examples of their recently perfected 'majolica' at the 1851 exhibition. This ware bore no resemblance to the similarly named 15th- and 16th-century Italian *maiolica*; the latter was a coarse buff pottery coated with an opaque white glaze that could be painted in colours, whereas the Victorian was a white pottery decorated with coloured transparent glazes. If an early parallel was required it would have been more appropriate to have named it after the 16th-century French potter, Bernard Palissy, whose work was similar. It differed from the colour-glazed articles made by Josiah Wedgwood when he was working with Thomas Whieldon a century earlier only because fresh colours had become available. They had been introduced in the first instance in 1850 by the French art director at Mintons, Léon Arnoux, but ten years passed before Wedgwood's marketed their own colour-glazed majolicas.

Llewellynn Jewitt wrote of the introduction of majolica at Etruria, stating that its principal exponent was Emile Lessore. He fled from France in 1858, and after a few months with Mintons transferred to their rivals. Lessore had a highly individual style, had spent many years as a painter in oils and successfully mastered the different technique of the medium he adopted. He employed normal opaque enamels and, in fact, his works were much closer to 15th- and 16th-century Italian *maiolica* than to the very different majolica. Lessore's style of painting was quite dissimilar to that of the average china-painter, which is more akin to the miniaturists; his slight yet assured draughtsmanship often recalls some of the pen and ink sketches of the Tiepolos, especially in the depiction of *putti*. He invariably signed his work.

Emile Lessore remained at Etruria until 1863, when he decided that he could no longer resist returning to his native land. There, he continued to decorate Wedgwood ware, which was sent across the Channel to him and was duly returned to Staffordshire. He went to live at Marlotte, near Fontainbleau, where during the siege of Paris 'many of his finest works were concealed by him in the cellars of his cottage'. Lessore died in 1876 at the age of 71.

Various fashions in printed ware came and went over the years. The middle shade of blue was replaced by a ghostly pale blue, which was rivalled in popularity by brown, green, raspberry red and others. Then, in the middle of the 1830s came a liking for a blurred dark blue, named 'flowing blue' or 'flow blue', achieved by the use of a 'volatilising agent' in the kiln causing the printed lines to spread and flow in the glaze. It was still being made in the 1870s.

Other types of ware that had a fashion in their day include those made in imitation of Saint Porchaire or Henri II pottery. This, too, was following in the footsteps of Mintons, where their employee Charles Toft made close copies of the original pieces. The cream or white clay object was decorated by having a pattern impressed in the soft surface, the cavities being filled with stained clays of which the surplus was removed. Wedgwood's made a chess table in the style that, it was said, was for use with John Flaxman's chessmen: they must have formed a remarkable combination. Another novelty was the use of photography in decoration. Jewitt wrote in 1878 that 'this forms a notable feature of progress in scientific decoration, and it is only meet that as photography itself was the undoubted discovery of a Wedgwood, its development as an aid to ceramic decoration should be left to his successors at the present day'.

Local pride in their most distinguished citizen led in 1859 to the forming of a committee to supervise the erection of a statue to commemorate Josiah Wedgwood, the statue to be paid for by public subscription. The sculptor chosen was Edward Davis, who had trained at the Royal Academy and worked in the studio of E. H. Baily, the latter remembered for his figure of Nelson standing on the top of the Column in Trafalgar Square, London. Wedgwood was represented with a Portland Vase in his hand, and once the work had gained approval it was cast in bronze. On 24 February 1863, in the presence of civic personalities, the clergy and general public, as well as the local Volunteers with a band and cannon, the Earl of Harrowby performed the unveiling ceremony. The statue still stands where it was first placed, in front of Stoke-on-Trent railway station.

Another project to the same end was also slowly coming to fruition: a Wedgwood Memorial Institute to be sited in Burslem. The idea of this had preceded the intention to have a statue, but being a much larger scheme it required greater and more sustained effort. Eight months after the formalities at Stoke the first stone of the Wedgwood Institute was laid, the ceremony being performed on 26 October 1863 by one of the most popular public figures of the day, the Rt. Hon. W. E. Gladstone. It was an indication of the regard in which Wedgwood was still held that no less a political figure than the Chancellor of the Exchequer should travel from London for the occasion, and that he should follow the brief stone-laying by a speech of considerable length. In it, he gave a few facts about the life of the potter, a task made difficult by the absence of printed details with the exception of those recorded half a century earlier by Simeon Shaw. Gladstone referred to this unfortunate lack of information:

Surely it is strange that the life of such a man, should, in this 'nation of shopkeepers,' yet at this date remain unwritten; and I have heard with pleasure a rumour, which I trust is true, that such a gap in our literature is about to be filled up.

In 1865 the gap was doubly filled: by Llewellynn Jewitt and Eliza Meteyard. Jewitt rushed to get his book published first, but he had been unable to gain access to the vitally important documents owned by Joseph Mayer and had to manage without them; the result was in the nature of *Hamlet* without the prince. The documents had been promised to Miss Meteyard, who made good use of them, dedicating her work, as did Jewitt, to Gladstone.

After so many decades in decline or merely static, the Wedgwood

85. Queensware jug painted by Emile Lessore. Impressed 'WEDGWOOD' and with date marks for August 1864. Height 21.2 cm. (8⅜ in.). City Museum & Art Gallery, Stoke-on-Trent.

company began to recover some of the enterprise they had enjoyed in the days of old Josiah. Charles Toft left Mintons for Etruria to become chief modeller (1872–89). From the same factory in 1876 came Thomas Allen, who became chief designer and art director at Etruria until he retired at the age of seventy-four in 1905. In the years he spent with Mintons, Allen had assiduously imitated the work of 18th-century Sèvres decorators on vases and other articles just as closely copied from those of the royal manufactory. At Wedgwood's he was not restricted to the 18th-century style, and his signed works included a striking series of pseudo-Elizabethan characters from Shakespeare painted on circular pottery plaques. Each of them has an unmistakable 1880s look in spite of costume details of earlier date.

The firm initiated two important changes: in 1875 a London showroom was again acquired, and three years later the making of bone-china

86. Wine (*left*) and Water (*right*) vases of bronzed basaltes modelled from plaster casts supplied by Flaxman in 1775. Impressed 'WEDGWOOD', *c.* 1900. Height 40.6 cm. (16 in.). Buten Museum of Wedgwood, Merion, Pennsylvania.

recommenced to continue uninterruptedly to the present day. The showroom at Holborn Circus, on the western fringe of the City, was at first shared with a manufacturer of medical glassware who also represented Wedgwood's. Ten year's later, the latter had the place to themselves, and in 1890 a move was made to another address in the same area: to 108 Hatton Garden, a street more generally associated with the diamond trade. In 1911 a further removal took place to 26–7 Hatton Garden, where the firm remained until the early 1940s.

In 1906 a Wedgwood Museum was opened at Etruria, sufficient interesting and historic documents and pottery having been assembled despite the loss of so much at the dispersal of the contents of the York Street premises in 1828–9. A catalogue was compiled by the London dealer and Wedgwood specialist, Frederick Rathbone, who provided brief descriptions of the 448 exhibits as well as introductions to the various types of ware on display and a chapter on the marks used by Josiah and his successors.

Although there was a steady flow of new shapes and decorative patterns, Wedgwood's continued to rely largely on the reissue of old and loved patterns. These were brought up to date by the alteration of

87. Blue and white jasper plaque of dancing girls and children designed by Charles Toft in 1887; impressed 'Wedgwood'. Width 23.5 cm. (9¼ in.). Buten Museum of Wedgwood, Merion, Pennsylvania.

88. Jasper part-teaset ornamented with relief patterns on a dotted gold ground, *c.* 1900; impressed 'WEDGWOOD ENGLAND'. Width of teapot 17.1 cm. (6¾ in.). City Museum & Art Gallery, Stoke-on-Trent.

89. Queensware figure of a polar bear designed by John Skeaping. Impressed
'J SKEAPING WEDGWOOD', *c.* 1930. Height 17.8 cm. (7 in.). City Museum &
Art Gallery, Stoke-on-Trent.

90. Queensware plaque painted by Thomas Allen with a named portrait of
Sir John Falstaff. Impressed 'WEDGWOOD' and with date marks for April
1879. Diameter 36.8 cm. (14½ in.). Wedgwood Museum, Barlaston.

such details as handles and knobs, suited to the taste of the time.

A new member was added to the design staff in 1912, Daisy Makeig-Jones, who in due course initiated a series of wares that broke completely with the Wedgwood tradition with regard to shapes, patterns and colours. She created a strange world inhabited by imaginary folk

91. Two covered vases decorated by Louise Powell. Both painted with the artist's monogram and impressed 'WEDGWOOD', one also with 'MADE IN ENGLAND', c. 1910. Height 50.8 cm. (20 in.). City Museum & Art Gallery, Stoke-on-Trent.

92. (*Left*) Slip-decorated jasper match-box and cover by Harry Barnard, who held various positions with Wedgwood's between 1896 and 1932. Impressed 'WEDGWOOD ENGLAND' and with Barnard's signature incised, *c.* 1920. Length 9.5 cm. (3¾ in.). City Museum & Art Gallery, Stoke-on-Trent.
93. (*Right*) Preserves pot and cover with black printed decoration, designed by Eric Ravilious *c.* 1930. Printed inscription 'Designed by Eric Ravilious'. Height 13.2 cm. (5⅕ in.). Wedgwood Museum, Barlaston.

copied from the works of Edmund Dulac and other book-illustrators, using fairies and goblins, dragons, butterflies and much else. They were printed in gold outline on vividly coloured lustred grounds on bone-china, the vases and other objects being in shapes reminiscent of those of 18th-century Chinese porcelain. Nothing could possibly have been more distant from the neo-classical simplicity of Josiah's day, but the British public of the 1920s liked and bought it.

At the close of the decade, when there was a world-wide recession following the stock market crash on Wall Street, Wedgwood's countered by attempting some innovations. They followed their founder's formula by commissioning work from leading artists and modellers, including John Skeaping, Eric Ravilious and Keith Murray. The first-named modelled a series of highly stylised animals; Ravilious engraved some interesting patterns for a revival of black overglaze printing; and Keith Murray designed some shapes that would have earned the approval of the first Josiah. In jasper, a dark olive-green and a rich crimson were introduced, the latter for the second time, but were withdrawn from production when it was found that the colours had a tendency to stain the white reliefs.

In addition to the difficult trading conditions then prevailing, another problem urgently demanded attention: Etruria was proving inadequate for modern manufacturing methods and could not be enlarged as other premises had been erected in its proximity. Also, the building was slowly and steadily subsiding, so that it became sited no less than eight

94, 95. Queensware plate, front and back, transfer-printed in sepia under the glaze. Diameter 20.3 cm. (8 in.).

feet below the level of the canal running alongside it. It was decided that a move should be made to Barlaston, five miles distant to the south-east, where a new factory and village would be built. This duly took place, the first part of the new building being opened in 1940, with the remainder gradually taking shape and being fully occupied ten years later. Etruria,

96. Bone-china vase, *c.* 1920, decorated in colours, gold and lustre (Fairyland Lustre) with the Candlemas pattern designed by Daisy Mackeig-Jones. 'WEDGWOOD ENGLAND' printed in gold and pattern number 'z5157'. Height 20.3 cm. (8 in.). Wedgwood Museum, Barlaston.

which had been in use for a total of 181 years, stood ghost-like and empty, being finally demolished in 1966. At some time in the 1950s the original plaster model of the statue of Josiah erected outside Stoke railway station was discovered in a store. A further bronze was cast from it, and this now stands facing the office entrance at Barlaston; the master-potter watching day and night over the firm he founded.

Since moving into their new premises, Wedgwood's have continued making the wares for which they have for so long been famous, continually adding to them as is essential in a competitive world. Among the products that have become increasingly popular are articles suitably inscribed in commemoration of some old or new, public or private, occasion. This is no more than a continuation of old Josiah Wedgwood's 'First Day Vases', his 'Prince of Wales Vase' or his Bastille medallion and other pieces referring to events in his lifetime. A similar continuity is exemplified in the present-day making of wares in 18th-century patterns, employing for the purpose moulds from the original blocks. There could be no better tribute than this to the far-seeing man who started it all.

97. Queensware vase designed by Keith Murray, c. 1935. Impressed 'WEDGWOOD' with traces of 'MADE IN ENGLAND' and printed with a facsimile of the artist's signature. Height 18.5 cm. (7¼ in.). City Museum & Art Gallery, Stoke-on-Trent.

# CHAPTER SIX

# Collecting Wedgwood Wares

THE WEDGWOOD collector is considerably aided by the fact that most genuine specimens are unequivocally recognisable. Early in his career Josiah Wedgwood realised that to mark his products in a particular way would gain the public's confidence and help to establish his reputation. In England at that time only a few of the porcelain-makers sporadically marked their goods with a sign of their own, or copied the well-known crossed swords of Meissen, but the pottery manufacturers had done no such thing. As early as 1759 Wedgwood began to put his name to his wares, and from 1769, the year when Etruria was opened, the productions of that manufactory were stamped with the names of the partners. The fact that the marks were impressed into the clay body by means of a metal stamp meant that they were difficult to remove; the use of acid or a grindstone for the purpose leaving a tell-tale depression. The effectiveness of Wedgwood's marks can be gauged by the existence of imitations that appeared in his lifetime, and caused him much anxiety.

The simple word 'Wedgwood' was used continuously from the start and can obviously confuse the unwary tyro. From 1891 it was given the addition of ENGLAND, and then MADE IN ENGLAND, which are obvious helps in dating, but to distinguish between basaltes and jasper made in the 18th century and that of the Victorian era requires experience. It is best to look upon a mark on any kind of chinaware as no more than a confirmation of other signs, and to consider these with care. Most collectors have the laudable ambition of possessing as many genuine early examples as possible, and it may be of assistance to them if the characteristics of the best pieces are considered.

For basaltes, jasper and other 'dry bodies' such as caneware and rosso antico they are as follows:

The most obvious sign is the high quality of finish, with every detail rendered clearly even when a relief is on a small scale. In successive decades the sharpness is likely to have diminished. Where a portion of a relief stands out it would have been undercut; work that was executed by hand much more painstakingly in the 18th century than at later dates.

The old pottery bodies were freed from impurities so far as was possible, and finished plaques and vases have much smoother surfaces than later specimens. To the finger an old piece has been described as having a 'wax-like' feel, 'which has the resistance of velvet without being woolly', whereas later ones are quite different to the touch and the white reliefs have a dry chalky appearance. The late Harry M. Buten considered that the smooth surface was acquired with age, akin to the patina of antique furniture, and he wrote: 'Soft velvety smoothness, featuring the patina that develops from two hundred years of polishing used to be important. Today, polishing jasper a few minutes with 400A wet or dry Trimite paper can reduce centuries of abrasion to a few minutes. Black paste shoe polish also helps to improve black basalt.' Not everyone shares the modern American passion for 're-finishing' anything old, and many prefer not to interfere with Nature.

Creamware is likewise distinguishable as regards old and new:

The earliest, which is seldom marked, is a rich, almost buff, colour and is thickly glazed.

Eighteenth-century marked pieces are noticeably light in the hand, the glaze is thin and evenly spread and the general finish neat. Later examples are heavier in weight, piece for piece, often thicker and less carefully finished. It is sometimes easy to overlook an impressed mark on a piece of glazed creamware, as the glaze may have run into the stamped letters with the result that they are semi-concealed.

More than one writer had averred that old Josiah Wedgwood used to appear in the factory from time to time inspecting the goods being made. If something did not come up to his standard, it was said, he would smash it with his walking stick or throw it on the floor saying: 'This is not good enough for Josiah.' Whether there is truth or not in the story is beside the point, but it has led to the supposition that everything that left Etruria was absolutely perfect in every particular. However, there is evidence of the opposite having taken place, and that Wedgwood was a realistic businessman who did not tolerate waste of goods that might be sold. When he was in London early in 1769 vase fever was raging; he wrote to his wife at Burslem regarding a wood carver, John Coward:

I have settled a plan & method with Mr. Coward to tinker all the black Vases that are crooked, we knock off the feet and fix wood ones, black'd, to them, those with snakes want$^g$. are to be supply$^d$. in the same way. I wish you could send me a parc$^l$. of these Invalids by Sundays Waggon, as he wishes to have w$^t$. we can furnish him w$^{th}$. of each sort together.

Later in the same year, in a letter to the London showroom, he mentioned creamware 'seconds', of which an increasing quantity was accumulating at Etruria. Wedgwood suggested that they might be disposed of in London, as they 'will never sell here at all, nor go along with the 2nd Table service we make up for Liverpool'. This is a clear indication that inferior ware was sent across the sea regularly, and confirming a worthwhile source of income that was not neglected. Wedgwood had referred to the subject in 1766 when discussing the drawback of making articles emblazoned with arms or crests. 'Plain ware,' he wrote, 'if it should not happen to be firsts, you will take off my hands as seconds, which, if Crested, would be ... useless ...'

Forgeries of Wedgwood wares exist and can sometimes be very deceptive, many of them being as old as their originals and bearing the characteristic signs of age. The Sèvres factory made some good versions of jasper plaques in the late 19th century, but the majority of Continental-made 'jasper' is of more recent date and is likely to deceive only beginners. It differs from Wedgwood in being poorly finished and, above all, in having been moulded in one piece, relief and all, with the background colour applied by brush round the outlines of the raised portion.

The more dangerous of the copies are those that were made in Staffordshire at the time when Wedgwood was making the prototypes. In particular, the copies of seals made by John Voyez, a modeller who worked for a short time at Etruria and then went to prison for an unspecified crime. Somehow he obtained suitable black clay, modelled it himself into passable imitations of basaltes seals and then, after marking each one with the names of Wedgwood and Bentley, paid a potter to fire them in a kiln. When he was asked why he put those names on his own seals he was reported as having had the effrontery to reply: 'I borrow & lend with them when I am out of any particular sorts, or they want any that I have, we borrow & lend with each other.' The work of Voyez is hard to detect, but his contemporaries were less daring in their piracies, putting their own, or no, name to their work. Among the marks used was one employed on seals which looks at a glance as if it is authentic, but closer inspection reveals that the name in minute lettering is WADG-WOJD.

98. Two black stoneware seals with designs in intaglio. Impressed 'WADGWOJD'. Widths 2.7 cm. ($1\frac{1}{16}$ in.) and 2.9 cm ($1\frac{1}{8}$ in.). The mark could be mistaken for that of Wedgwood.

99. Creamware dish with purple painted husk and flower decoration. Made
by C. J. Poskotchine at Morje, near St Petersburg (Leningrad), *c.* 1840.
A replacement for the service supplied to Catherine of Russia by
Josiah Wedgwood in 1770. Width 33 cm. (13 in.). Wedgwood Museum,
Barlaston.

Creamware was no less extensively copied than the other Wedgwood
productions. It was made at a number of English factories, notably at
Leeds, Liverpool and Swansea where, in each case, marks were only
used occasionally. Across the Channel, a number of French factories
competed with English importations, in some instances employing
Staffordshire potters to assist them. Other factories were active in the
Netherlands, Germany, Scandinavia, Hungary, Italy, Spain and Russia.
In the latter country, a pottery at Morje, near St Petersburg, was directed
by C. J. Poskotchine between 1817 and 1847, then by Theodor Emel-
yanov. Poskotchine made some creamware replacements for the 'husk'
service that had been supplied to Catherine the Great by Wedgwood in
1770.

Much of the ware made on the mainland of Europe might be mistaken
for English, but a good proportion is appropriately marked to denote
its origin. The fact that a specimen is unmarked does not prevent its
source being suspect by reason of its shape, decoration and general
colour. Donald Towner has pointed out that these creamwares have
been the subject of comparatively little study, and makes a start in a brief
chapter by detailing some of their characteristics. He includes those of
the French *faïence fine* or *faïence anglaise*, the German *steingut*, the Dutch
*Engels steen*, the Danish *stengods*, the Italian *terraglia inglese*, and the
Swedish *flintporslin*; all of which owed their success to the industry of
Josiah Wedgwood.

# *Wedgwood Marks*

(Impressed, unless otherwise stated)

WEDGWOOD
Wedgwood          in various sizes, on creamware from *c.* 1769

            in relief on a thin disc or wafer 1769–80

Wedgwood          used on seals with the addition of the catalogue
& Bentley          number of the subject

W & B             used on seals of very small size

WEDGWOOD
& BENTLEY
                  in various sizes on ornamental articles, 1769–80
Wedgwood
& Bentley

                  in various sizes on ornamental and useful articles
WEDGWOOD          following the death of Bentley in 1780, and for an
Wedgwood          indeterminate period after Josiah Wedgwood's
                  death in 1795

WEDGWOOD & SONS   used briefly in 1790

JOSIAH            found only on some tripod incense burners and used
WEDGWOOD          for an unknown reason. Miss Meteyard recorded
Feb. 2nd. 1805    an example with the year reading '1085'

                  used on bone china *c.* 1812–29 printed or stencilled in
WEDGWOOD          red, black or gold; printed in blue on pieces
                  decorated in underglaze blue

WEDGWOOD'S
STONE CHINA

printed on stoneware *c.* 1827–61

WEDGWOOD
ETRURIA

used in various type faces for a short time, *c.* 1840–5

PEARL

on pearlware *c.* 1840–68

P

on pearlware after 1868

WEDGWOOD

used on bone china from 1878, printed in various colours. ENGLAND added from *c.* 1891, MADE IN ENGLAND from *c.* 1910

WEDGWOOD
Bone China
MADE IN ENGLAND

later version of the foregoing, printed

WEDGWOOD
MADE IN
ENGLAND

used on Queensware, printed in various colours, from *c.* 1940

From 1860 an additional impressed mark of three letters of the alphabet was brought into use, indicating the month and year of manufacture and the potter responsible for making the article. After 1930 the final letter, which denoted the year, was replaced by the last two numbers of the year in question: for example 32 = 1932. Many modern books listing marks on pottery and porcelain decode these letters and give their equivalents.

# Selected Bibliography

Harry Barnard, *Chats on Wedgwood*, London, 1924 (reprinted Merion, Pa., 1970).

Harry M. Buten, *Wedgwood Rarities*, Merion, Pa., 1969.

A. H. Church, *Josiah Wedgwood, Master Potter*, London, 1894.

K. E. Farrer (ed.), *Letters of Josiah Wedgwood 1762–80*, 2 vols., London, 1903. *Letters of Josiah Wedgwood 1781–95*, London, 1906 (all three reprinted in 3 vols. Manchester, 1973.

Ann Finer and George Savage (eds.), *Selected Letters of Josiah Wedgwood*, London, 1965.

J. K. des Fontaines (ed.) *Wedgwood's 1880 Catalogue of Shapes*, London, 1971.

Una des Fontaines, *Wedgwood Fairyland Lustre*, London, 1975.

M. H. Grant, *The Makers of Black Basaltes*, London, 1910 (reprinted 1970).

W. B. Honey, *Wedgwood Ware*, London, 1948.

Llewellynn Jewitt, *The Wedgwoods, being a Life of Josiah Wedgwood*, London, 1865. *The Ceramic Art of Great Britain*, 2 vols., London, 1878.

Alison Kelly, *Decorative Wedgwood*, London, 1965.

Chas. F. C. Luxmoore, '*Saltglaze*', *with The Notes of a Collector*, Exeter, 1924 (reprinted 1971).

Wolf Mankowitz, *The Portland Vase and the Wedgwood Copies*, London, 1952. *Wedgwood*, London, 1953.

Eliza Meteyard, *The Life of Josiah Wedgwood*, 2 vols., London, 1865–6 (reprinted 1970). *A Group of Englishmen*, London, 1871. *The Wedgwood Handbook*, London, 1875 (reprinted New York 1963).

Arnold R. Mountford, *Staffordshire Salt-glazed Stoneware*, London, 1971.

*Proceedings of The Wedgwood Society*, London, 1956–

Frederick Rathbone, *Old Wedgwood*, London, 1898 (reprinted Merion, Pa. 1968). *A Catalogue of the Wedgwood Museum, Etruria*, Etruria, 1909.

Robin Reilly and George Savage, *Wedgwood: The Portrait Medallions*, London, 1973.

Simeon Shaw, *History of the Staffordshire Potteries*, Hanley, 1829 (reprinted Newton Abbot 1970).

Samuel Smiles, *Josiah Wedgwood, F.R.S.*, London, 1894.

Donald Towner, *Creamware*, London, 1978.

John Ward (and Simeon Shaw), *The Borough of Stoke-upon-Trent*, London, 1843.

Josiah C. Wedgwood, *A History of the Wedgwood Family*, London, 1908.

## ACKNOWLEDGEMENTS

N.B. Illustration numbers are given.
Bearnes, Torquay 37; Buten Museum of Wedgwood, Merion, Pennsylvania 58, 69, 70, 80, 86, 87; Castle Museum, Nottingham 21, 23, 42, 46, 49, 53, 54, 60, 65; Christies, London 40, 51, 55; City Museum & Art Gallery, Plymouth 4; City Museum & Art Gallery, Stoke-on-Trent 1, 2, 7, 8, 15, 36, 64, 75, 81, 85, 88, 89, 91, 92, 97; County Museum, Truro 9, 10; Leeds Art Galleries 84; Mansell Collection, London 59; The National Trust: Saltram House, Devon 16; Dr and Mrs Ricks, Oklahoma 71 (*group*); Sotheby Parke Bernet & Co., London 12, 18, 35, 41; E. N. Stretton, 31; Victoria and Albert Museum, London 3, 6, 13, 14, 34, 56; The Wedgwood Museum, Barlaston *Frontispiece*, 11, 20, 22, 24, 25, 26, 27, 28, 29, 30, 32, 38, 44, 45, 47, 52, 61, 66, 67, 71 (*pedestal*), 73, 76, 77, 82, 83, 90, 93, 96, 99.

# Index